Multiple Choice Questions in Human Physiology

D0988367

ALSO AVAILABLE FROM ARNOLD

Multiple Choice Questions in Anatomy
by J Pegington 0 340 50785 3

Multiple Choice Questions in Anatomy for the FRCS
by J McDiarmid, J Bernard, T Greenwell, A Li, Nick Marshall, C Thurnell
and C Stone 0 340 59433 0

**Multiple Choice Questions in Applied Basic Sciences for the FRCS
Examinations**
by J Pegington, F Imms, D Davies and P Boulos 0 340 57317 1

Multiple Choice Questions in Clinical Pharmacology
by T Mant, L Lewis and J Ritter 0 340 55932 2

**Multiple Choice Questions in Clinical Pharmacology for the Higher Level
Examination**
by G McCarthy and R Mirakhur 0 340 55792 3

Multiple Choice Questions in Haematology, Second Edition
by T P Baglin, J E G Braithwaite and T R Mitchell 0 340 55303 0

Multiple Choice Questions in Gynaecology and Obstetrics, Third Edition
by M Setchell and R Lilford 0 340 58896 9

Multiple Choice Questions in Pathology, Third Edition,
by Ian Brown 0 340 55164 X

Multiple Choice Questions in Pathology for the FRCS
by J McDiarmid, J Bernard, T Greenwell, A Li, Nick Marshall, C Thurnell
and C Stone 0 340 59434 9

Multiple Choice Questions in Pharmacology, Second Edition
by A D'Mello and Z Kruk 0 340 54321 3

Multiple Choice Questions in Physiology, Third Edition
by L Bindman, P Ellaway, B Jewell, and L Smaje 0 340 67677 9

Multiple Choice Questions in Physiology for the FRCS
by J McDiarmid J Bernard, T Greenwell, A Li, Nick Marshall, C Thurnell
and C Stone 0 340 59435 7

Essays and MCQs in Anaesthesia and Intensive Care
by P Murphy 0 340 62524 4

MULTIPLE CHOICE QUESTIONS IN HUMAN PHYSIOLOGY

with answers and explanatory comments

Fifth Edition

Ian C Roddie CBE, DSc, MD, FRCPI

Emeritus Professor of Physiology, The Queen's University of Belfast;
former Head of Medical Education, National Guard King Khalid Hospital,
Jeddah, Saudi Arabia

and

William FM Wallace BSc, MD, FRCP, FRCA

Professor of Applied Physiology, The Queen's University of Belfast and
Consultant in Physiology, Belfast City Hospital, Belfast, N. Ireland

A member of the Hodder Headline Group
LONDON • SYDNEY • AUCKLAND
Co-published in the USA by Oxford University Press, Inc., New York

First published in Great Britain 1991
Fifth edition published in 1997 –
Second impression 1997 by Arnold,
a member of the Hodder Headline Group
338 Euston Road, London NW1 3BH

Co-published in the United States of America by
Oxford University Press, Inc.,
198 Madison Avenue, New York, NY 10016
Oxford is a registered trademark of Oxford University Press

British Library Cataloguing in Publication Data
A catalogue record for this book is available from the British Library

Library of Congress Cataloging-in-Publication Data
A catalog record for this book is available from the Library of Congress

ISBN 0 340 66234 4

Produced and typeset in 8/10 pt Helvetica by Gray Publishing, Tunbridge Wells
Printed and bound in Great Britain by J W Arrowsmith Ltd, Bristol

CONTENTS

PREFACE

This book has now reached its fifth edition since it was first published some 25 years ago. Our aim to base the questions on generally accepted aspects of physiology most relevant to clinical practice seems to have been fulfilled – medical, dental and other health care students and doctors in specialty training in countries around the world have told us of the book's relevance and usefulness.

We have tried to cover most of the concepts and knowledge typically asked for in physiology examinations and to concentrate on the core knowledge that is essential to pass them. We believe that students who score consistently well in these questions know enough to face most examinations in physiology with confidence. By concentrating on the area where yes/no answers can be given to questions with reasonable certainty, we have had to exclude areas where knowledge is as yet conjectural and speculative. We have tried to avoid excessive detail in the way of facts and figures; those which are included are of value in medical practice. Both conventional and SI units are generally quoted. Comments on the answers are given on the page opposite each question. We hope that, with the comments, the book will provide a compact revision tutor, encouraging understanding rather than rote learning.

For most questions the common five-branch MCQ format has been used. The stem and a single branch constitute a statement to be judged True or False by the reader. Care has been taken that the statements in any question are not mutually exclusive, so five independent decisions are required to answer each question. This system has the advantage of simplicity and brevity over most other forms of multiple choice question. The questions are grouped in 13 sections covering the various physiological systems. They cover both basic and applied aspects of the subject. The applied questions are designed so that the answers may be deduced mainly by making use of basic physiological knowledge and should provide a link with clinical practice.

In this edition, a further opportunity has been taken to prune and edit questions for greater compactness, clarity and precision and to bring in new areas of knowledge which have emerged since the last edition went to press. A new section on sports and exercise physiology has been added since this is an area of growing importance. We have also added a new section with 'interpretative' questions to provide practice in the interpretation of data. This section contains some diagram-based questions; the number is small since such questions are expensive in page space. The final section contains some examples of extended matching questions, an alternative form of multiple choice question where answers have to be selected from lists of options. In these questions each option can be used once, more than once or not at all.

We thank colleagues for suggesting questions and all who commented on previous editions. We continue to welcome such comments.

ICR
WFMW

HOW TO USE THIS BOOK

1 *A stimulus to fill gaps in your knowledge*

This book is intended as a revision tutor and should help you to revise your physiology in preparation for examinations. It is particularly aimed at helping you to identify areas where your knowledge and understanding need to be improved. The statements in this book are presented so that you can commit yourself in written opinion and can then confirm correct information and identify errors. The comments should reinforce your knowledge when you were correct and indicate why you were mistaken if your answer was wrong.

2 *Scoring your answers*

(a) Answer, say, 20 questions (100 decisions), aiming to complete them in about 50 minutes. In our experience of this type of question (one point tested in each part), it is best for candidates to answer virtually all questions.

(b) Score your answers by giving +1 for a correct response, −1 for an incorrect response and 0 for any omitted. It is suggested that this approach is in line with professional life when many true/false decisions must be taken – send the patient to hospital? Begin a certain treatment? Carry out surgery urgently? The penalties for a wrong decision can be considerable!

(c) As a very approximate guide, the following scale would apply to candidates who have not spent time memorizing particular questions:

50–60	fair
60–70	good
70–90	excellent
90–100	outstanding

1 BODY FLUIDS

1 Extracellular fluid differs from intracellular fluid in that its:
   (a) Volume is greater
   (b) Tonicity is lower
   (c) Anions are mainly inorganic
   (d) Sodium:potassium molar ratio is higher
   (e) pH is lower

2 Blood group antigens (agglutinogens) are:
   (a) Carried on the haemoglobin molecule
   (b) Beta globulins
   (c) Equally immunogenic
   (d) Present in fetal blood
   (e) Inherited as recessive Mendelian characteristics

3 Total body water, expressed as a percentage of body weight:
   (a) Can be measured with an indicator dilution technique using deuterium oxide
   (b) Is smaller on average in women than in men
   (c) Rises following injection of posterior pituitary extracts
   (d) Falls during starvation
   (e) Is less than 80% in young adults

4 Breakdown of erythrocytes in the body:
   (a) Occurs when they are 6–8 weeks old
   (b) Takes place in the reticulo-endothelial system
   (c) Yields iron, most of which is excreted in the urine
   (d) Yields bilirubin which is carried by plasma protein to the liver
   (e) Is required for the synthesis of bile salts

5 A person with group A blood:
   (a) Has anti-B antibody in the plasma
   (b) May have the genotype AB
   (c) May have a parent with group O blood
   (d) May have children with group A or group O blood only
   (e) Whose partner is also A can only have children of group A or O

6 Blood platelets assist in arresting bleeding by:
   (a) Releasing factors promoting blood clotting
   (b) Adhering together to form plugs when exposed to collagen
   (c) Liberating high concentrations of calcium
   (d) Releasing factors causing vasoconstriction
   (e) Inhibiting fibrinolysis by blocking the conversion of plasminogen to plasmin

1 **(a) False** Cells contain half to two-thirds of the total body fluid.
  (b) False It is the same; if it were lower, osmosis would draw water into the cells.
  (c) True Mainly Cl^- and HCO_3^-; inside, the main anions are protein and organic phosphates.
  (d) True Around 30:1; the intracellular ratio is about 1:10.
  (e) False Intracellular pH is lower due to cellular metabolism.

2 **(a) False** They are part of the red cell membrane.
  (b) False They are glycoproteins.
  (c) False A, B and D are more immunogenic than the others.
  (d) True Fetal blood may elicit immune responses if it enters the maternal circulation.
  (e) False They are Mendelian dominants.

3 **(a) True** D_2O (heavy water) exchanges with water in all body fluid compartments.
  (b) True Women carry more fat than men and fat has a low water content.
  (c) True ADH in the extracts inhibits water excretion by the kidneys.
  (d) False It rises as fat stores are metabolized to provide energy.
  (e) True 70%, the percentage in the lean body mass, is about the maximum percentage possible.

4 **(a) False** The normal erythrocyte life-span is 16–18 weeks.
  (b) True The RES removes effete RBCs from the circulation.
  (c) False Most of the iron is retained for further use.
  (d) True The protein makes the bilirubin relatively water soluble.
  (e) False Bile salts are synthesized from sterols in the liver.

5 **(a) True** This appears about the time of birth.
  (b) False This would make them blood group AB.
  (c) True They could inherit an A gene from the other parent to give genotype AO.
  (d) False B or AB is possible depending on the partner's genes.
  (e) True In this case, neither parent has the B gene.

6 **(a) True** For example, thromboplastin, part of the intrinsic pathway.
  (b) True Vascular leaks are sealed by such platelet plugs.
  (c) False High Ca^{2+} levels are not needed for haemostasis; normal levels are adequate.
  (d) True For example, serotonin (5-hydroxytryptamine).
  (e) False Serotonin from platelets can release vascular plasminogen activators.

7 Plasma bilirubin:
   (a) Is a steroid pigment
   (b) Is converted to biliverdin in the liver
   (c) Does not normally cross cerebral capillary walls
   (d) Is freely filtered in the renal glomerulus
   (e) Is sensitive to light

8 Monocytes:
   (a) Originate from precursor cells in lymph nodes
   (b) Can increase in number when their parent cells are stimulated by factors released from activated lymphocytes
   (c) Unlike granulocytes, do not migrate across capillary walls
   (d) Can transform into large multinucleated cells in certain chronic infections
   (e) Manufacture immunoglobulin M

9 Erythrocytes:
   (a) Are responsible for the major part of blood viscosity
   (b) Contain the enzyme carbonic anhydrase
   (c) Metabolize glucose to produce CO_2 and H_2O
   (d) Swell to bursting point when suspended in 0.9% (150 mmol/litre) saline
   (e) Have rigid walls

10 Human plasma albumin:
   (a) Contributes more to plasma colloid osmotic pressure than globulin
   (b) Filters freely at the renal glomerulus
   (c) Is negatively charged at the normal pH of blood
   (d) Carries carbon dioxide in blood
   (e) Lacks the essential amino acids

11 Neutrophil granulocytes:
   (a) Are the most common leucocyte in normal blood
   (b) Contain proteolytic enzymes
   (c) Have a life-span in the circulation of 3–4 weeks
   (d) Contain actin and myosin microfilaments
   (e) Are present in high concentration in pus

12 Bleeding from a small cut in the skin:
   (a) Is normally diminished by local vascular spasm
   (b) Ceases within about 5 min in normal people
   (c) Is prolonged in severe factor VIII (antihaemophilic globulin) deficiency
   (d) Is greater from warm skin than from cold skin
   (e) Is reduced if the affected limb is elevated

7 (a) False It is a porphyrin pigment derived from haem.
(b) False Bilirubin is derived from biliverdin formed from haem, not the other way about.
(c) True The blood–brain barrier normally prevents bilirubin entering brain tissue.
(d) False The bilirubin–protein complex is too large to pass the glomerular filter.
(e) True Light converts bilirubin to lumirubin which is excreted more rapidly; phototherapy may be used in the treatment of haemolytic jaundice in children.

8 (a) False They originate from stem cells in bone marrow.
(b) True Activated T cells release GMCSF (granulocyte/macrophage colony stimulating factor) which stimulates monocyte stem cells to proliferate.
(c) False After 4–6 days in the circulation, monocytes migrate out to become tissue macrophages.
(d) True The 'giant cells' seen in tissues affected by tuberculosis and leprosy.
(e) False Immunoglobulins are made by ribosomes in lymphocytes.

9 (a) True Blood viscosity rises exponentially with the haematocrit.
(b) True It catalyses the reaction $CO_2 + H_2O \rightleftharpoons H^+ + HCO_3^-$.
(c) True Glycolysis generates the energy needed to maintain electrochemical gradients across their membranes.
(d) False This is isotonic with their contents.
(e) False The walls deform easily to squeeze through capillaries.

10 (a) True Its greater mass and lower molecular weight provide more osmotically active particles.
(b) False Only a small amount is filtered normally and this is reabsorbed by the tubules.
(c) True Blood pH is well above albumin's isoelectric point so negative charges (COO^-) predominate.
(d) True As carbamino protein ($R\text{-}NH_2 + CO_2 \rightleftharpoons R\text{-}NH.COOH$).
(e) False It is a first class protein containing essential and non-essential amino acids.

11 (a) True They comprise 60–70% of circulating leucocytes.
(b) True Their granules contain such enzymes which, with toxic oxygen metabolites, can kill and digest the bacteria they engulf.
(c) False Less than a day.
(d) True Responsible for their amoeboid motility.
(e) True Pus consists largely of dead neutrophils.

12 (a) True Due to the effects of tissue damage and serotonin on vascular smooth muscle.
(b) True This is the upper limit of the normal 'bleeding time'.
(c) False Factor VIII deficiency increases clotting time, not bleeding time.
(d) True Warmth dilates skin blood vessels.
(e) True Intravascular pressure is reduced in an elevated limb.

13 Antibodies:
  (a) Are protein molecules
  (b) Are absent from the blood in early fetal life
  (c) Are produced at a greater rate after a first, than after a second, exposure to an antigen 6 weeks later
  (d) Circulating as free immunoglobulins are produced by B lymphocytes
  (e) With a 1 in 8 titre are more concentrated than ones with a 1 in 4 titre

14 Circulating red blood cells:
  (a) Are about 1% nucleated
  (b) May show an intracellular network pattern if appropriately stained
  (c) Are distributed evenly across the blood stream in large blood vessels
  (d) Travel at slower velocity in venules than in capillaries
  (e) Deform as they pass through the capillaries

15 Lymphocytes:
  (a) Constitute 1–2% of circulating white cells
  (b) Are motile
  (c) Can transform into plasma cells
  (d) Decrease in number following removal of the adult thymus gland
  (e) Decrease in number during immunosuppressive drug therapy

16 The specific gravity (relative density) of:
  (a) Red cells is less than that of plasma
  (b) Plasma is due more to its protein than to its electrolyte content
  (c) Plasma decreases as extracellular fluid and electrolytes are lost
  (d) Blood is higher on average in women than in men
  (e) Urine can fall below 1.000 in a water diuresis

17 Blood:
  (a) Makes up about 7% of body weight
  (b) Forms a higher percentage of body weight in fat than in thin people
  (c) Volume can be calculated by multiplying plasma volume by the haematocrit (expressed as a percentage)
  (d) Volume rises after water is drunk
  (e) Expresses serum when it clots

18 The cell membranes in skeletal muscle:
  (a) Are impermeable to fat-soluble substances
  (b) Are more permeable to sodium than to potassium ions
  (c) Become more permeable to glucose in the presence of insulin
  (d) Become less permeable to potassium in the presence of insulin
  (e) Show invaginations which connect to a system of intracellular tubules involved in excitation–contraction coupling

13   **(a)**   **True**   They are made by ribosomes in plasma cells.
     (b)   **True**   Immunological tolerance prevents the fetus forming antibodies to its own proteins.
     (c)   **False**   The response to the second exposure is greater since the immune system has been sensitized by the first exposure.
     (d)   **True**   T lymphocytes are responsible for cell-mediated immunity.
     (e)   **True**   Antibody with a 1 in 8 titre is detected at greater dilution than one with a 1 in 4 titre.

14   **(a)**   **False**   Nucleated red cells are not normally seen in peripheral blood.
     (b)   **True**   Reticulocytes, the most immature circulating RBCs, show this pattern when stained with supravital stains such as brilliant cresyl blue.
     (c)   **False**   They form an axial stream away from the vessel wall.
     (d)   **False**   The capillary bed has a greater total cross-sectional area than the venular bed.
     (e)   **True**   Normal cells, around 7 μm in diameter, become bullet shaped as they pass through 5 μm diameter capillaries.

15   **(a)**   **False**   About 20% of leucocytes are lymphocytes.
     (b)   **True**   They migrate by amoeboid movement to areas of chronic inflammation.
     (c)   **True**   As plasma cells they manufacture humoral antibodies.
     (d)   **False**   The thymus is atrophied and has little function in the adult.
     (e)   **True**   Lymphocytes and immune responses are closely linked.

16   **(a)**   **False**   Red cells are heavier and hence sediment on standing.
     (b)   **True**   The mass of plasma proteins (70–80 g/litre) far exceeds that of plasma electrolytes (about 10 g/litre).
     (c)   **False**   It increases; plasma specific gravity is an index of ECF volume if protein levels are normal.
     (d)   **False**   It is higher in men, who have a higher haematocrit.
     (e)   **False**   The specific gravity of pure water is 1.000; urine is water plus solutes.

17   **(a)**   **True**   For example, 5 kg (about 5 litres) in a 70 kg man.
     (b)   **False**   Since fat tissue is relatively avascular, the reverse is true.
     (c)   **False**   It can be calculated by multiplying plasma volume by 1/1-haematocrit (expressed as a decimal).
     (d)   **True**   The water is absorbed into the blood.
     (e)   **True**   Serum is plasma minus its clotting factors.

18   **(a)**   **False**   The membrane consists largely of lipid.
     (b)   **False**   The reverse is true; sodium ions, being more hydrated than potassium ions, are larger complexes.
     (c)   **True**   Thus glucose is stored as muscle glycogen after a meal
     (d)   **False**   They become more permeable; injections of insulin and glucose lower the serum potassium level.
     (e)   **True**   These are called the T system of tubules.

19 The osmolality of:
   (a) A solution determines its freezing point
   (b) Intracellular fluid is about twice that of extracellular fluid
   (c) 1.8% sodium chloride is about twice that of normal plasma
   (d) 5% dextrose solution is about five times that of 0.9% saline
   (e) Plasma is due more to its protein than to its electrolyte content

20 The pH:
   (a) Of arterial blood normally ranges from 7.2 to 7.6
   (b) Units express $[H^+]$ in moles/litre
   (c) Of blood is directly proportional to the P_{CO_2}
   (d) Of blood is directly proportional to $[HCO_3^-]$
   (e) Of urine is normally less than 7

21 Cerebrospinal fluid:
   (a) Is an ultrafiltrate of plasma
   (b) Is the main source of the brain's nutrition
   (c) Has the same pH as arterial blood
   (d) Has a higher glucose concentration than has plasma
   (e) Has a higher calcium concentration than has plasma

22 Antigens:
   (a) Are usually proteins or polypeptide molecules
   (b) Can only be recognized by immune system cells previously exposed to that antigen
   (c) Are normally absorbed from the gut via lymphatics and carried to mesenteric lymph nodes
   (d) Induce a smaller immune response when protein synthesis is suppressed
   (e) Are taken up by antigen-presenting macrophages which activate the immune system

23 Blood eosinophils:
   (a) Have agranular cytoplasm
   (b) Are about a quarter of all leucocytes
   (c) Are relatively abundant in the mucosa of the respiratory, urinary and alimentary tracts
   (d) Release cytokines
   (e) Increase in number in viral infections

24 Normal blood clotting requires:
   (a) Inactivation of heparin
   (b) Inactivation of plasmin (fibrinolysin)
   (c) Calcium ions
   (d) An adequate intake of vitamin K
   (e) An adequate intake of vitamin C

19 (a) **True** Depression of the freezing point is an index of a solution's osmolality.
   (b) **False** Their osmolality is the same; osmotic water movements ensure that this is so.
   (c) **True** Plasma has the tonicity of a normal saline solution (0.9% sodium chloride).
   (d) **False** They have the same number of particles.
   (e) **False** Proteins account for only 1% of plasma osmolality.

20 (a) **False** The range is normally between 7.35 and 7.45.
   (b) **False** They express it as the negative logarithm of the $[H^+]$ in moles/litre.
   (c) **True** P_{CO_2} determines the carbonic acid concentration $[H_2CO_3]$.
   (d) **False** It is inversely proportional to $[HCO_3^-]$.
   (e) **True** The normal diet leaves acidic, rather than alkaline, residues.

21 (a) **False** It is secreted actively by the choroid plexuses.
   (b) **False** Brain nutrition is delivered mainly by cerebral blood flow.
   (c) **False** It is around 7.3 compared with 7.4 in blood.
   (d) **False** It is about two-thirds that of plasma.
   (e) **False** About half; protein-bound calcium is negligible in CSF.

22 (a) **True** Large carbohydrate molecules may also be antigenic.
   (b) **False** The ability to recognize foreign antigens is innate and does not depend on previous exposure to them.
   (c) **False** Antigens, being proteins or carbohydrates, are not normally absorbed; they are digested in the gut.
   (d) **True** Antibodies are proteins synthesized by ribosomes in activated lymphocytes.
   (e) **True** Antigens can also act directly on receptors on lymphocyte membranes.

23 (a) **False** They have eosinophilic granules (eosinophilic granulocytes).
   (b) **False** Only 1–4% of white cells are eosinophils.
   (c) **True** They are involved in mucosal immunity.
   (d) **True** Interleukin 4 and platelet activating factor (PAF).
   (e) **False** Their number increases in parasitic infections and allergic conditions.

24 (a) **False** The anticoagulant effects of heparin are overwhelmed.
   (c) **False** Blood clots in spite of the fibrinolytic system.
   (b) **True** Removal of calcium ions prevents clotting.
   (d) **True** Vitamin K is needed by the liver for synthesis of prothrombin and other factors.
   (e) **False** The spontaneous bleeding from the gums, etc., seen in scurvy is due to capillary abnormality, not a clotting defect.

25 Antibodies (agglutinins) of the A and B red cell antigens (agglutinogens):
    (a) Are present in fetal plasma
    (b) Cause haemolysis of RBCs containing the A and B antigens when added
        to blood
    (c) Do not normally cross the placental barrier
    (d) Have a molecular weight in excess of 500 000
    (e) Are monovalent

26 Lymph:
    (a) Contains plasma proteins
    (b) Vessels are involved in the absorption of amino acids from the intestine
    (c) Production increases during muscular activity
    (d) Does not normally contain cells
    (e) Flow is aided by contraction of adjacent skeletal muscles

27 Blood platelets:
    (a) Are formed in the bone marrow
    (b) Are normally more numerous than white cells
    (c) Have a small single-lobed nucleus
    (d) Increase in number after injury and surgery
    (e) Alter shape when in contact with collagen

28 The conversion of fibrinogen to fibrin:
    (a) Is effected by prothrombin
    (b) Involves the disruption of certain peptide linkages by a proteolytic enzyme
    (c) Is followed by polymerization of fibrin monomers
    (d) Is inhibited by heparin
    (e) Is reversed by plasmin (fibrinolysin)

29 An appropriate dilution indicator for measuring:
    (a) Total body water is sucrose
    (b) Plasma volume is radioactive sodium
    (c) Extracellular fluid volume is inulin
    (d) Intracellular fluid volume directly is heavy water (deuterium oxide)
    (e) Total body potassium is radioactive potassium

25 (a) **False** They form shortly after birth, possibly in response to A and B antigens carried into the body by invading bacteria.
    (b) **False** They cause agglutination (clumping) of A, B and AB cells.
    (c) **True**   Unlike Rh antibodies which have a smaller molecular size.
    (d) **True**   Around 1 000 000.
    (e) **False** They are divalent and hence cause red cells to adhere to one another during agglutination.

26 (a) **True**   Derived from plasma proteins leaked from capillaries into the tissues; it returns these to the blood
    (b) **False** Lymph vessels are involved in the uptake and transport of absorbed fat.
    (c) **True**   Increased capillary pressure due to muscle vasodilation increases tissue fluid formation.
    (d) **False** It contains lymphocytes derived from lymph nodes.
    (e) **True**   In addition, intrinsic rhythmic contractions in lymphatics help to propel lymph.

27 (a) **True**   They are formed from megalokaryocytes.
    (b) **True**   By a factor of 20 or more.
    (c) **False** No nucleus, but the cytoplasm contains electron-dense granules, lysosomes and mitochondria..
    (d) **True**   This increases the tendency of blood to clot.
    (e) **True**   They put out pseudopodia and adhere to the collagen and to one another.

28 (a) **False** It is effected by thrombin; prothombin is the inactive precursor of thrombin.
    (b) **True**   Thrombin breaks off the solubilizing end-groups.
    (c) **True**   Polymerized fibrin monomers form the strands of the clot meshwork.
    (d) **True**   This is a rapidly acting anticoagulant.
    (e) **False** Plasmin does not convert fibrin back to fibrinogen, it degrades both fibrin and fibrinogen to products which can inhibit thrombin.

29 (a) **False** Sucrose does not cross the cell membrane freely to equilibrate with ICF.
    (b) **False** Sodium ions migrate easily from plasma to equilibrate with interstitial fluid.
    (c) **True**   Inulin crosses capillary walls freely but does not enter cells.
    (d) **False** ICF volume is not measured directly; it is calculated by measuring ECF volume and total body water and subtracting the former from the latter.
    (e) **True**   Radioactive K^+ equilibrates with the body pool of non-radioactive K^+; both isotopes are treated similarly in the body.

30 Thirst can be:
    (a) Produced by a rise in plasma tonicity
    (b) Produced by stimulation of certain areas in the hypothalamus
    (c) Produced by a fall in blood volume
    (d) Associated with decreased secretion of ADH
    (e) Relieved by water intake before the water has been absorbed from the gut

31 Intravenous infusion of:
    (a) Two litres of normal saline restores blood volume in a patient who
         suddenly lost 2 litres of blood
    (b) Bicarbonate is appropriate for patients being treated for cardiac and
         respiratory arrest
    (c) Potassium-free fluids is appropriate for a patient with severe vomiting
    (d) Isotonic glucose will expand both intracellular and extracellular fluid
         compartments
    (e) Hypertonic saline will raise intracellular osmolality

32 Excessive tissue fluid (oedema) in the legs may:
    (a) Be associated with a raised extracellular fluid volume
    (b) Result from hepatic disease
    (c) Result from blockage of pelvic lymphatics
    (d) Increase local interstitial fluid pressure
    (e) Result from a high arterial blood pressure in the absence of heart failure

33 Haemolytic disease of the new-born:
    (a) Affects mainly babies of Rh-positive mothers
    (b) Occurs mainly in babies who lack D agglutinogen
    (c) Causes jaundice which clears rapidly after birth
    (d) Can be treated by transfusing the affected baby with Rh-positive blood
    (e) Can be prevented by injecting the mother with anti-D agglutinins just after
         delivery

34 The appearance of centrifuged blood may suggest that:
    (a) Anaemia is present if there is more plasma than packed cells
    (b) The plasma lipid level is high
    (c) The patient has jaundice
    (d) Haemolysis has occurred
    (e) The patient has leukaemia

30 **(a) True**   Stimulation of osmoreceptors by the increased tonicity generates thirst sensation.

(b) True   The supraoptic nucleus of the hypothalamus contains osmoreceptors.

(c) True   This can happen, even though blood tonicity is unchanged; volume receptors may be involved.

(d) False   ADH secretion is increased.

(e) True   Flushing out the mouth with water can provide temporary relief from thirst.

31 **(a) False**   Some of the saline escapes from the circulation to the interstitial fluid.

(b) True   It corrects the acidosis caused by accumulation of lactic acid and CO_2 in the tissues.

(c) False   Alimentary secretions are rich in potassium.

(d) True   Glucose is metabolized, leaving the water to be distributed in both compartments

(e) True   Hypertonic extracellular fluid will draw water osmotically from the cells.

32 **(a) True**   Oedema is an increase in the interstitial component of ECF.

(b) True   Albumin deficiency reduces plasma colloid osmotic pressure.

(c) True   Protein accumulates in interstitial fluid and reduces the colloid osmotic pressure gradient across the capillary wall.

(d) True   This contributes to a new pressure equilibrium.

(e) False   Arteriolar constriction in hypertension raises arterial, but not capillary, pressure.

33 **(a) False**   It affects babies of Rh-negative mothers.

(b) False   It occurs in Rh-positive babies.

(c) False   The jaundice deepens rapidly after birth as bilirubin is no longer excreted by the maternal liver.

(d) False   This would be attacked by maternal Rh antibodies in the infant's blood; Rh-negative blood is given.

(e) True   These destroy fetal Rh-positive cells in the maternal circulation before they sensitize her to D antigen.

34 **(a) False**   The normal percentage of plasma in centrifuged blood is about 55%.

(b) True   The plasma is cloudy or even milky.

(c) True   The plasma is yellow.

(d) True   The plasma is red.

(e) True   The buffy coat is greatly thickened.

35 Patients with moderate to severe anaemia have a reduced:
    (a) Cardiac output
    (b) Incidence of vascular bruits
    (c) 2:3-diphosphoglycerate blood level
    (d) Arterial P_{O_2}
    (e) Capacity to raise oxygen consumption in exercise

36 Iron deficiency:
    (a) Frequently follows persistent loss of blood from the body
    (b) Is more common in men than in women
    (c) May cause anaemia by inhibiting the rate of multiplication of RBC stem cells
    (d) May cause large pale erythrocytes to appear in peripheral blood
    (e) Anaemia should normally be treated by injections of iron

37 Severe reactions are likely after transfusion of blood group:
    (a) A to a group B person
    (b) O to a group AB person
    (c) A to a group O person
    (d) A to a group AB person
    (e) O Rh-negative to a group AB Rh-positive person

38 The haematocrit (packed cell volume):
    (a) May be obtained by centrifugation of blood
    (b) May be calculated by multiplying the mean cell volume by the red cell count
    (c) Rises in a patient who sustains widespread burns
    (d) Rises following injections of aldosterone
    (e) Rises in macrocytic megaloblastic anaemias such as pernicious (B_{12} deficiency) anaemia

39 Red cell formation is increased:
    (a) By giving vitamin B_{12} injections to healthy people on a normal diet
    (b) In blood donors 1 week after a blood donation
    (c) In patients with haemolytic anaemia
    (d) By giving injections of erythropoietin to nephrectomized patients
    (e) In patients who have a raised blood reticulocyte count

35 (a) False Output rises to compensate for the blood's reduced O_2-carrying capacity.

  (b) False Bruits are common since increased flow velocity and decreased blood viscosity increase the likelihood of turbulent flow.

  (c) False 2:3-DPG is increased, shifting the dissociation curve to the right so that blood gives up its oxygen more easily.

  (d) False Arterial P_{O_2} is normal; it is O_2 content which is reduced.

  (e) True Due to the reduced capacity to deliver O_2 to the muscles.

36 (a) True Especially if dietary intake of iron is limited.

  (b) False It is more common in women due to menstrual blood loss.

  (c) False It causes anaemia by limiting the rate of haemoglobin synthesis.

  (d) False In iron deficiency anaemia, RBCs are small and pale due to lack of haemoglobin.

  (e) False Oral iron is avidly absorbed in iron deficiency states.

37 (a) True The recipients have anti-A antibody.

  (b) False Group O people are 'universal donors'.

  (c) True The recipients have anti-A antibody.

  (d) False Group AB persons, 'universal recipients', lack anti-A and anti-B antibodies.

  (e) False The recipients lack anti-A, anti-B and anti-Rh antibodies.

38 (a) True Since red cells are heavier than plasma.

  (b) True This gives a slightly lower value than centrifugation which traps a little plasma between cells.

  (c) True Due to loss of plasma and interstitial fluid.

  (d) False It falls as extracellular fluid and hence plasma volume increases.

  (e) False Though individual RBCs are large, total red cell mass is decreased.

39 (a) False Healthy normal people do not benefit from vitamin B_{12} supplements.

  (b) True The RBC deficit is corrected by bone marrow stimulation by erythropoietin.

  (c) True The reduced oxygen-carrying capacity of the blood causes release of erythropoietin which stimulates RBC stem cells in the bone marrow.

  (d) True The anaemia seen in nephrectomized patients is due largely to lack of erythropoietin.

  (e) True A raised reticulocyte count is evidence of a hyperactive bone marrow.

40   Vitamin B_{12} deficiency may:
    (a)   Result from disease of the terminal part of the ileum
    (b)   Result in anaemia with small RBCs well filled with haemoglobin
    (c)   Cause wasting (atrophy) of the gastric mucosa
    (d)   Cause a reduction in the circulating platelet level
    (e)   Cause pathological changes in the central nervous system

41   A raised blood pH and bicarbonate level is consistent with:
    (a)   Metabolic acidosis
    (b)   Partly compensated respiratory alkalosis
    (c)   A reduced P_{CO_2}
    (d)   Chronic renal failure with a raised P_{CO_2}
    (e)   A history of persistent vomiting of gastric contents

42   A patient with partly compensated respiratory acidosis:
    (a)   Must have a raised P_{CO_2}
    (b)   May have a reduced hydrogen ion concentration [H^+]
    (c)   Must have a raised bicarbonate concentration [HCO_3^-]
    (d)   May have evidence of renal compensation
    (e)   May have respiratory failure due to hypoventilation

43   A patient with an uncompensated respiratory alkalosis may have:
    (a)   Been exposed to living at high altitudes
    (b)   A reduced [H_2CO_3]:[HCO_3^-] ratio
    (c)   Neuromuscular hyperexcitability
    (d)   An arterial pH of 7.3
    (e)   A blood [H^+] of 30 nmol/litre

44   In investigating a patient's acid–base status:
    (a)   Venous rather than arterial blood should be studied
    (b)   Blood samples may be stored for up to 12 h at room temperature before analysis
    (c)   pH can be calculated if [HCO_3^-] and P_{CO_2} are known
    (d)   Raised urinary ammonium salts suggest renal compensation for respiratory acidosis
    (e)   An early fall in [HCO_3^-] suggests that the acid–base disturbance is respiratory in origin

45   Respiratory alkalosis differs from metabolic alkalosis in that the:
    (a)   Likelihood of tetany is less
    (b)   Urine is alkaline
    (c)   [HCO_3^-] is normal or low
    (d) P_{CO_2} is reduced
    (e)   Reduction in cerebral blood flow is greater

40   **(a)**   **True**   The B_{12}/intrinsic factor complex is absorbed in the terminal ileum.

     (b)   **False**   Lack of B_{12} results in a macrocytic hyperchromic anaemia.

     (c)   **False**   Gastric mucosa atrophy is the cause, not the effect, of B_{12} lack; gastric mucosa normally produces the 'intrinsic factor' required for B_{12} absorption.

     (d)   **True** B_{12} is used in the DNA synthesis required by platelet precursor cells.

     (e)   **True**   Maintenance of myelin in neural sheaths also depends on vitamin B_{12}.

41   **(a)**   **False**   It is consistent with a metabolic alkalosis.

     (b)   **False**   A partly compensated acidosis has a low pH.

     (c)   **False** P_{CO2} is normally raised in metabolic alkalosis as a compensatory mechanism.

     (d)   **False**   All these values are reduced in chronic renal failure.

     (e)   **True**   Pyloric obstruction causes a metabolic alkalosis.

42   **(a)**   **True**   This is the hallmark of a respiratory acidosis.

     (b)   **False** $[H^+]$ is raised in uncompensated acidosis.

     (c)   **True**   The raised $[HCO_3^-]$ is compensating partly for the raised P_{CO_2}.

     (d)   **True**   The raised $[HCO_3^-]$, compensating the raised P_{CO_2}, is generated by the kidneys.

     (e)   **True**   This leads to retention of carbon dioxide.

43   **(a)**   **True**   This is consistent with hypoxia-induced hyperventilation.

     (b)   **True**   This is consistent with alkalosis.

     (c)   **True**   Alkalosis favours the development of tetany by increasing the binding power of plasma protein for ionic calcium.

     (d)   **False**   This is an acidotic pH.

     (e)   **True**   The normal level is 40 nmol/litre.

44   **(a)**   **False**   Only arterial blood is precisely regulated for $[H^+]$.

     (b)   **False**   Analysis should be prompt; acid–base status is affected by blood cell metabolism.

     (c)   **True**   pH is a function of their ratio.

     (d)   **True**   Ammonia is secreted to buffer the hydrogen ions being excreted as the kidneys manufacture bicarbonate.

     (e)   **False**   A primary respiratory acid–base problem leads initially to an altered P_{CO_2}.

45   **(a)**   **False**   Both kinds of alkalosis may result in tetany.

     (b)   **False**   It is likely to be alkaline in both.

     (c)   **True** $[HCO_3^-]$ is raised in metabolic alkalosis but falls to compensate for the low P_{CO_2} in respiratory alkalosis.

     (d)   **True** P_{CO_2} is reduced in respiratory alkalosis but rises to compensate for the high $[H_2CO_3^-]$ in metabolic alkalosis.

     (e)   **True**   The greater fall in P_{CO_2} in respiratory alkalosis causes more cerebral vasoconstriction.

46 Rejection of a transplanted organ is made less likely by:
    (a) Treatment which reduces the blood lymphocyte count
    (b) Keeping the recipient in a germ-free environment
    (c) Irradiation of the transplanted organ with X-rays
    (d) Drugs which interfere with mitosis
    (e) Transplanting between identical twins

47 Reduction in the neutrophil granulocyte count may be:
    (a) Caused by drugs suppressing bone marrow activity
    (b) A consequence of tissue damage
    (c) Associated with painful throat ulcers
    (d) Associated with widespread purulent infections
    (e) Caused by high levels of circulating glucocorticoids

48 A fall in plasma sodium concentration:
    (a) May result from excessive production of ADH
    (b) Decreases intracellular fluid volume
    (c) May occur in people engaged in hard physical work in humid tropical
         climates
    (d) Reduces plasma osmolality
    (e) Is likely to cause thirst

49 Sodium retention:
    (a) Occurs for several days after major surgery
    (b) Occurs in response to secretion of aldosterone, but not cortisol
    (c) Expands the extracellular fluid volume
    (d) Expands the blood volume
    (e) Increases the severity of oedema

50 Sodium depletion differs from sodium retention in that it causes a reduction in:
    (a) Central venous pressure
    (b) Renin production
    (c) The specific gravity of the blood
    (d) Intracellular fluid volume
    (e) Total body mass

51 Sodium depletion differs from water depletion in that:
    (a) Cardiovascular changes are less pronounced
    (b) Intracellular fluid volume is less affected
    (c) The haematocrit increases
    (d) Thirst is more severe
    (e) Antidiuretic hormone levels are higher

46 **(a) True**   T lymphocytes are responsible for tissue rejection.
   (b) False   This environment may be necessary because of suppression of the recipient's immune responses; it has no bearing on the rejection process.
   (c) False   This would not affect the transplant antigens.
   (d) True   These suppress the multiplication of lymphocytic stem cells.
   (e) True   Identical twins have identical antigens and do not reject each other's tissues.

47 **(a) True**   Granulocytes are formed in the bone marrow.
   (b) False   Production of neutrophils increases following tissue damage.
   (c) True   Neutrophils are not available to kill bacterial invaders.
   (d) False   There will not be much pus since pus consists mainly of dead neutrophils.
   (e) False   These suppress lymphocytes and eosinophils.

48 **(a) True**   Due to excessive reabsorption of water from the collecting ducts of the nephron.
   (b) False   Water is drawn into cells from the hypotonic extracellular fluid; water intoxication may occur.
   (c) True   People sweating heavily may replace their water, but not their salt, deficit; they tend to get muscle cramps unless they supplement their salt intake.
   (d) True   Sodium ions are responsible for nearly half of plasma osmolality.
   (e) False   The hypothalamic osmoreceptors responsible for thirst respond to hypertonicity, not hypotonicity of the ECF.

49 **(a) True**   This is part of the metabolic response to trauma.
   (b) False   Both have mineralocorticoid effects.
   (c) True   Sodium chloride is the 'skeleton' of the ECF; chloride and water are retained with the sodium.
   (d) True   Plasma is part of extracellular volume.
   (e) True   Oedema fluid is excess interstitial fluid.

50 **(a) True**   Blood volume parallels body sodium levels; it expands with sodium retention and shrinks with sodium depletion.
   (b) False   A reduced blood volume stimulates release of renin.
   (c) False   It is increased in sodium depletion due to an increased haematocrit.
   (d) False   If anything, ICF volume expands osmotically in sodium depletion.
   (e) True   Due to the loss of extracellular fluid in sodium depletion.

51 **(a) False**   Blood volume is more reduced with sodium depletion; cardiovascular changes are more pronounced.
   (b) True   Extracellular volume is a function of body sodium content.
   (c) False   It increases in both cases.
   (d) False   Hypertonicity is the main stimulus causing thirst.
   (e) False   Here also, hypertonicity is the main stimulus for ADH secretion.

52 Potassium depletion:
    (a) Can be detected by analysis of a biopsied sample of muscle
    (b) Can result from loss of gastrointestinal secretions
    (c) Causes increased activity of intestinal smooth muscle
    (d) Exacerbates pre-existing acidosis
    (e) Increases T wave amplitude in the electrocardiogram

53 A high blood potassium level (hyperkalaemia):
    (a) Occurs in acute renal failure
    (b) Follows severe crush injuries to the limbs
    (c) May diminish cardiac performance and cause death
    (d) Increases skeletal muscle strength
    (e) May be reduced by intravenous infusion of insulin and glucose

54 Deficiency of factor VIII (antihaemophilic globulin):
    (a) Increases the bleeding time
    (b) Is due to an abnormal gene on the Y chromosome
    (c) To 75% of its normal value results in excessive bleeding after tooth
         extraction
    (d) Causes small (petechial) haemorrhages into the skin to cause purpura
    (e) Affects the extrinsic, rather than the intrinsic, pathway for blood
         coagulation

55 A raised level of calcium in the blood (hypercalcaemia):
    (a) May occur when parathyroid activity decreases
    (b) May occur when the plasma protein level falls
    (c) May occur in chronic renal failure
    (d) Causes increased excitability of nerve and muscle
    (e) Increases the risk of stone formation in the urinary tract

56 Intravenous infusion of 1 litre of:
    (a) Normal (isotonic) saline increases the ECF more than the ICF volume
    (b) 10% dextrose provides sufficient energy for a sedentary adult for one day
    (c) A suspension of lipids provides 2–3 times the energy of a suspension of
         carbohydrates with the same concentration
    (d) Isotonic (5%) dextrose raises total body water by 1–5% in the average
         adult
    (e) An amino acid solution provides 3–4 times the energy of a carbohydrate
         solution with the same concentration

52 (a) **True**   Since most body potassium is intracellular.
   (b) **True**   Gastrointestinal secretions are rich in potassium.
   (c) **False** Activity decreases and intestinal paralysis may occur.
   (d) **False** K^+ competes with H^+ for excretion in the renal tubules; a low $[K^+]$ favours renal excretion of H^+ ions and this would reduce the severity of acidosis.
   (e) **False** The amplitude of the T waves decreases.

53 (a) **True**   Due to inability to excrete K^+ ingested and released from cell breakdown in the body.
   (b) **True**   Potassium is released from the damaged muscle fibres.
   (c) **True**   Abnormal rhythms and heart failure may result.
   (d) **False** Both hypokalaemia and hyperkalaemia cause skeletal muscle weakness.
   (e) **True**   This facilitates entry of potassium into cells.

54 (a) **False** Clotting time is increased, but bleeding time is determined by platelets and by vascular contraction.
   (b) **False** It is due to a recessive abnormality of the X chromosome.
   (c) **False** Abnormal bleeding does not occur until the level falls below 50%.
   (d) **False** Purpura is caused by capillary or platelet disorders.
   (e) **False** It affects the intrinsic pathway.

55 (a) **False** This reduces blood calcium.
   (b) **False** This lowers the bound, and hence the total, calcium level.
   (c) **False** In chronic renal failure PO_4 retention raises blood PO_4 levels; Ca^{2+} levels fall to maintain a constant $[Ca^{2+}]\,[PO_4^-]$ product.
   (d) **False** It depresses excitability.
   (e) **True**   More calcium is filtered and this increases the urinary $[Ca^{2+}]\,[PO_4^-]$ solubility product.

56 (a) **True**   Sodium and chloride remain mainly extracellular.
   (b) **False** It provide less than a quarter of the daily energy requirement.
   (c) **True**   1 g of fat when oxidized liberates 2–3 times the energy liberated by 1 g of carbohydrate.
   (d) **True**   Total body water (about 40 litres) increase to about 41 litres (2.5% increase).
   (e) **False** Amino acids and carbohydrates provide similar energy per unit weight but amino acids are useful for maintaining body tissue proteins.

57 Myocardial blood flow to the left ventricle increases during:
   (a) Early systole
   (b) Myocardial hypoxia
   (c) Adenosine infusions
   (d) Stimulation of sympathetic nerves to the heart
   (e) Arterial hypertension

58 Local metabolic activity is the chief factor determining the rate of blood flow to:
   (a) Heart
   (b) Skin
   (c) Skeletal muscle
   (d) Lung
   (e) Kidney

59 The pressure:
   (a) Drop along large veins is similar to that along large arteries
   (b) Drop across the hepatic portal bed is similar to that across the splenic vascular bed
   (c) In the hepatic portal vein exceeds that in the inferior vena cava
   (d) Drop across the vascular bed in the foot is greater when a subject is in the vertical than when he is in the horizontal position
   (e) In foot veins is lower when walking than when standing still

60 The second heart sound differs from the first heart sound in that it is:
   (a) Related to turbulence set up by valve closure
   (b) Longer lasting than the first sound
   (c) Higher in frequency
   (d) Occasionally split
   (e) Heard when the ventricles are relaxing

61 Pulmonary vascular resistance is:
   (a) Less than one-third that offered by the systemic circuit
   (b) Decreased when alveolar oxygen pressure falls
   (c) Expressed in units of volume flow per unit time per unit pressure gradient
   (d) Decreased during exercise
   (e) Regulated reflexly to regulate the rate of pulmonary blood flow

57 (a) **False** It falls; coronary vessels are compressed by the contracting myocardium.
   (b) **True** A fall in P_{O_2} has a potent vasodilator effect on coronary vessels.
   (c) **True** Adenosine released from hypoxic myocardium is also a potent vasodilator.
   (d) **True** Sympathetic stimulation increases the rate and force of contraction; the resulting increase in the rate of production of vasodilator metabolites dilates coronary vessels.
   (e) **True** Myocardial work and metabolism are increased in hypertension.

58 (a) **True** There is a close relationship between the work of the heart and coronary flow.
   (b) **False** Skin blood flow is geared mainly to thermoregulation and normally exceeds that needed for the skin's modest metabolic requirements.
   (c) **True** However, in severe exercise, an oxygen debt may develop as metabolic requirements outstrip the capacity of the circulation to deliver oxygen.
   (d) **False** Pulmonary blood flow is determined by prevailing cardiac output.
   (e) **False** As in skin, renal blood flow normally exceeds local metabolic needs.

59 (a) **True** About 10 mmHg or less; both offer little resistance to flow.
   (b) **False** The drop across the splenic vascular bed (about 60 mmHg) is much larger; the hepatic portal bed offers little resistance to flow.
   (c) **True** Otherwise blood would not flow through the portal bed.
   (d) **False** Changing from the horizontal to the vertical position increases arterial and venous pressures equally.
   (e) **True** The muscle pump in the leg decreases venous pressure.

60 (a) **False** This applies to both heart sounds.
   (b) **False** It is about 20% shorter.
   (c) **True** About 50 Hz compared with 35 Hz for the first sound.
   (d) **False** Both may be split due to asynchronous valve closure.
   (e) **True** The first sound is due to ventricular systole; the second occurs during ventricular relaxation when the aortic valves snap shut as ventricular pressure falls below aortic.

61 (a) **True** The pressure head needed to drive cardiac output through the pulmonary circuit (about 15 mmHg) is much less than that needed in the systemic circuit (about 90 mmHg).
   (b) **False** The reverse is true; low alveolar P_{O_2} may cause pulmonary hypertension.
   (c) **False** These are conductance units, the reciprocal of resistance units.
   (d) **True** Thus there is little rise in pulmonary arterial pressure. Release of nitric oxide from the pulmonary vascular endothelium may account for the vasodilatation.
   (e) **False** Pulmonary blood flow must equal left ventricular (cardiac) output.

62   Ventricular filling:
     (a)   Depends mainly on atrial contraction
     (b)   Begins during isometric ventricular relaxation
     (c)   Gives rise to a third heart sound in some healthy people
     (d)   Can occur only when atrial pressure is greater than atmospheric pressure
     (e)   Is most rapid in the first half of diastole

63   Veins:
     (a)   Contain most of the blood volume
     (b)   Have a sympathetic vasoconstrictor innervation
     (c)   Receive nutrition from vasa vasorum arising from their lumen
     (d)   Respond to distension by contraction of their smooth muscle
     (e)   Undergo smooth muscle hypertrophy when exposed to arterial pressure
           through an arteriovenous fistula

64   In the heart:
     (a)   The left atrial wall is about three times thicker than the right atrial wall
     (b)   Systolic contraction normally begins in the left atrium
     (c)   Excitation spreads directly from atrial to ventricular fibres
     (d)   Atrial and ventricular muscle contracts simultaneously in systole
     (e)   The contracting ventricles shorten from apex to base

65   Isometric (static) exercise differs from isotonic (dynamic) exercise in that it
     causes a greater increase in:
     (a)   Venous return
     (b)   Pressure in the veins draining the exercising muscle
     (c)   Muscle blood flow
     (d)   Mean arterial pressure
     (e)   Cardiac work for the same increase in cardiac output

66   The net loss of fluid from capillaries in the legs is increased by:
     (a)   Arteriolar dilation
     (b)   Change from the recumbent to the standing position
     (c)   Lymphatic obstruction
     (d)   Leg exercise
     (e)   Plasma albumin depletion

62 **(a) False** Atrial contraction accounts for only about 20% of filling at rest.
    (b) False During this phase the AV valves are closed.
    (c) True This low-pitched sound is sometimes heard in early diastole.
    (d) False Filling occurs when atrial pressure exceeds ventricular pressure.
    (e) True Due to entry of blood accumulated in the atria during ventricular systole.

63 **(a) True** Around three-quarters; veins are referred to as 'capacitance' vessels.
    (b) True These regulate venous capacity.
    (c) False Their vasa arise from neighbouring arteries.
    (d) True This 'myogenic' response helps to limit the degree of distension.
    (e) True Another functional adaptation to resist distension.

64 **(a) False** Their wall thickness is similar since the work-load of the two atria is similar.
    (b) False It begins at the sinoatrial node in the right atrium.
    (c) False Excitation can only pass from atria to ventricles via specialized conducting tissue in the AV bundle.
    (d) False Delay of excitation in the AV bundle makes atrial precede ventricular contraction.
    (e) True Due to the spiral arrangement of some muscle fibres; circular fibres reduce ventricular circumference.

65 **(a) False** The muscle pump is more effective in dynamic than in static exercise.
    (b) True In dynamic exercise, the muscle pump decreases venous pressure.
    (c) False The increase is less since there is sustained compression of their blood supply.
    (d) True There is relatively little fall in total peripheral resistance with static exercise.
    (e) True The rise in arterial pressure with static exercise increases cardiac work since cardiac output has to be ejected against a higher aortic pressure.

66 **(a) True** This increases capillary hydrostatic pressure.
    (b) True In the standing position, capillary pressure increases by the hydrostatic equivalent of the column of blood below the heart.
    (c) False Lymphatic obstruction allows tissue fluid to accumulate; the rise in interstitial pressure reduces the capillary transmural hydrostatic pressure gradient.
    (d) True Capillary pressure rises during the exercise hyperaemia.
    (e) True Hypoproteinaemia decreases the transmural colloid osmotic pressure gradient.

67   When measuring blood pressure by the auscultatory method:
     (a) The sounds that are heard are generated in the heart
     (b) The cuff pressure at which the first sounds are heard indicates systolic
          pressure
     (c) The cuff pressure at which the loudest sounds are heard indicates
          diastolic pressure
     (d) Systolic pressure estimations tend to be lower than those made by the
          palpatory method
     (e) Wider cuffs are required for larger arms

68   The absolute refractory period in the ventricles:
     (a) Is the period when the ventricles are completely inexcitable
     (b) Corresponds to the period of ventricular depolarization
     (c) Corresponds to the period of ventricular contraction
     (d) Is shorter than the corresponding period in atrial muscle
     (e) Decreases during sympathetic stimulation of the heart

69   Vascular resistance increases:
     (a) By 50% when the vascular radius is halved
     (b) When wall thickness increases
     (c) When vessel length increases
     (d) When blood viscosity increases
     (e) From the arteriolar to the capillary bed

70   Sympathetic drive to the heart is increased:
     (a) In exercise
     (b) In excitement
     (c) In hypotension
     (d) When parasympathetic drive is decreased
     (e) During a vasovagal attack

71   In an adult subject standing quietly at rest, venous pressure in the:
     (a) Foot is approximately equal to arterial pressure at heart level
     (b) Thorax decreases when the subject inhales
     (c) Hand is subatmospheric when the hand is raised above the head
     (d) Venous sinuses of the skull is subatmospheric
     (e) Superior vena cava is an index of cardiac filling pressure

67 (a) **False** Korotkoff sounds are produced locally by the turbulence due to blood being forced through the partially occluded artery.
   (b) **True**   The sharp taps of Phase 1 are generated as the systolic pressure peaks force blood under the cuff.
   (c) **False** Sudden muffling (Phase 4) or disappearance (Phase 5) of the sounds indicates the diastolic pressure point.
   (d) **False** They are usually higher since palpation may fail to detect the first tiny pulses.
   (e) **True**   Otherwise the full cuff pressure may not be transmitted to the artery.

68 (a) **True**   This is the definition.
   (b) **True**   Depolarized cells cannot be excited.
   (c) **True**   This prevents tetanic contraction of the heart.
   (d) **False** It is longer, as is the duration of its depolarization.
   (e) **True**   Shortening the refractory period permits higher heart rates.

69 (a) **False** It increases 16-fold; resistance is related to the fourth power of the radius.
   (b) **False** Vascular resistance is not related to wall thickness.
   (c) **True**   It is directly proportional to length.
   (d) **True**   Resistance is related to Viscosity × Length/ Radius^4.
   (e) **False** Total arteriolar resistance exceeds total capillary resistance, though the reverse is true for single vessels.

70 (a) **True**   β-Adrenergic blocking drugs reduce exercise tachycardia.
   (b) **True**   The tachycardia during excitement is also reduced by ß-blocking drugs.
   (c) **True**   A reflex response to decreased drive from arterial baroreceptors.
   (d) **False** The two systems can function independently.
   (e) **False** Parasympathetic drive slows the heart during a vasovagal attack; sympathetic drive may stimulate the heart during recovery.

71 (a) **True**   It is about 90 mmHg, due to the column of blood (about 1 m) between the heart and the foot.
   (b) **True**   The negative intrathoracic pressure during inhalation is transmitted to the veins.
   (c) **False** Limb veins collapse above heart level so venous pressure is atmospheric.
   (d) **True**   The sinuses are held open by their meningeal attachments and cannot collapse.
   (e) **True**   A central venous pressure line is usually placed here.

72 Hyperaemia in skeletal muscle during exercise is normally associated with:
   (a) Release of sympathetic vasoconstrictor tone in the exercising muscles
   (b) Capillary dilation due to relaxation of capillary smooth muscle
   (c) A fall in arterial pressure
   (d) Reflex vasoconstriction in other vascular beds
   (e) An increase in cardiac output

73 Sinoatrial node cells are:
   (a) Found in both atria
   (b) Innervated by the vagus
   (c) Unable to generate impulses when completely denervated
   (d) Connected to the AV node by fine bundles of Purkinje tissue
   (e) Stable with regard to their membrane potentials

74 The first heart sound corresponds in time with:
   (a) Closure of the aortic and pulmonary valves
   (b) The P wave of the electrocardiogram
   (c) A rise in atrial pressure
   (d) A rise in ventricular pressure
   (e) The 'A' wave in central venous pressure

75 Increased sympathetic drive to the heart increases the:
   (a) Rate of diastolic depolarization in sinoatrial node cells
   (b) Coronary blood flow
   (c) Rate of conduction in Purkinje tissue
   (d) Slope of the Frank–Starling (work versus stretch) curve of the heart
   (e) Ejection fraction of the left ventricle

76 The velocity of blood flow:
   (a) In capillaries is low because they offer high resistance to flow
   (b) In veins is greater than in venules
   (c) Can fall to zero in the ascending aorta during diastole
   (d) Is greater towards the centre of large blood vessels than at the periphery
   (e) In the circulation falls as the haematocrit falls

72 (a) **False** Exercise hyperaemia occurs normally in sympathectomized muscles.

(b) **False** Capillaries have no smooth muscle; they dilate passively with the rise in capillary pressure due to active arteriolar dilation.

(c) **False** Arterial pressure usually rises.

(d) **True** Vasoconstriction in kidneys, gut and skin prevent excessive falls in total peripheral resistance.

(e) **True** This, together with the reflex constriction above, compensates for the fall in muscular vascular resistance and thus prevents arterial pressure falling during exercise.

73 (a) **False** The SA node is in the right atrium near its junction with the superior vena cava.

(b) **True** Vagal activity slows the rate of impulse generation and thus the heart rate.

(c) **False** The SA node has intrinsic rhythmicity and can generate impulses independently.

(d) **False** Purkinje tissue is confined to the ventricles; atrial fibres conduct impulses from the SA to the AV node.

(e) **False** They are unstable; impulse generation is due to spontaneous diastolic depolarization of the cells.

74 (a) **False** It is synchronous with mitral and tricuspid closure.

(b) **False** It corresponds with the QRS complex.

(c) **True** The mitral and tricuspid valves bulge into the atria.

(d) **True** This closes the mitral and tricuspid valves.

(e) **False** It corresponds with the C wave; the A wave is due to atrial contraction which precedes the first heart sound.

75 (a) **True** This increases the rate of impulse generation and hence heart rate.

(b) **True** The increase in myocardial metabolism generates vasodilator metabolites.

(c) **True** Rapid spread of excitation in the ventricles results in more forceful contractions as the ventricular fibres are activated nearly simultaneously.

(d) **True** This enhances the force of contraction at any given filling pressure.

(e) **True** Due to the increased force of contraction.

76 (a) **False** It is low because the capillary bed has a large total cross-sectional area.

(b) **True** The venous bed has a smaller total cross-sectional area than the venular bed.

(c) **True** There is a brief period of retrograde flow as the aortic valve closes.

(d) **True** Axial flow occurs in large vessels; near the walls, flow velocity is zero.

(e) **False** It rises due to the compensatory increase in cardiac output.

77 The strength of contraction of left ventricular muscle increases when:
    (a) End-diastolic ventricular filling pressure rises
    (b) Serum potassium levels rise
    (c) Blood calcium levels fall
    (d) Strenuous exercise is undertaken
    (e) Peripheral resistance is increased

78 During isometric ventricular contraction:
    (a) The entry and exit valves of the ventricle are closed
    (b) Pressure in the aorta rises
    (c) Pressure in the atria falls
    (d) Left coronary blood flow falls
    (e) The rate of rise in pressure is greater in the right than in the left ventricle

79 In the electrocardiogram, the:
    (a) QRS complex follows the onset of ventricular contraction
    (b) T wave is due to repolarization of the ventricles
    (c) PR interval corresponds with atrial depolarization
    (d) RT interval is related to ventricular action potential duration
    (e) R–R interval normally varies during the respiratory cycle

80 Cardiac output:
    (a) Is normally expressed as the output of one ventricle in litres/min
    (b) May not increase when heart rate rises
    (c) Usually rises when a person lies down
    (d) Rises in a hot environment
    (e) Does not increase in exercise following denervation of the heart

81 Arterioles:
    (a) Have a smaller wall:lumen ratio than have arteries
    (b) Play a major role in regulating arterial blood pressure
    (c) Play a major role in regulating local blood flow
    (d) Offer more resistance to flow than capillaries
    (e) Have a larger total cross-sectional area than do the capillaries

82 The Purkinje tissue cells in the heart:
    (a) Conduct impulses faster than some neurones
    (b) Are larger than ventricular myocardial cells
    (c) Lead to contraction of the base before the apex of the heart
    (d) Are responsible for the short duration of the QRS complex
    (e) Are responsible for the configuration of the QRS complex

77 (a) **True** As seen in the Frank–Starling left ventricular function curve.
    (b) **False** High K⁺ levels decrease cardiac contractility.
    (c) **False** Calcium channel blockers decrease cardiac contractility.
    (d) **True** Due to increased sympathetic drive to the ventricles and increased venous return.
    (e) **True** Initially by increased ventricular filling; later by ventricular hypertrophy.

78 (a) **True** Hence no blood leaves the ventricles.
    (b) **False** Pressure in the aorta falls as run-off of blood to the tissues continues.
    (c) **False** It rises as ventricular pressure causes the AV valves to bulge into the atria.
    (d) **True** The blood vessels are squeezed by the contracting myocardium.
    (e) **False** The greater force of contraction in the left ventricle gives a greater rate of rise of pressure.

79 (a) **False** The electrical event precedes the mechanical event.
    (b) **True** This electrical event corresponds with ventricular relaxation.
    (c) **False** It corresponds to the interval between atrial and ventricular depolarization due to delay of the impulse in the AV bundle.
    (d) **True** R indicates the beginning, and T the end, of the ventricular action potential.
    (e) **True** Due to the heart rate changes associated with respiratory sinus arrhythmia.

80 (a) **True** The output from the left and right ventricle is the same.
    (b) **True** It depends on what happens to stroke volume.
    (c) **True** Lying down normally increases the filling pressure of the heart.
    (d) **True** To meet the needs of increased metabolism and increased skin blood flow.
    (e) **False** The output does increase due to changes in the filling pressure, level of circulating hormones, etc.

81 (a) **False** In arterioles the ratio is much greater, at about 1:1.
    (b) **True** They provide most of the peripheral resistance.
    (c) **True** Local flow varies directly with the fourth power of their radii.
    (d) **True** The drop in pressure across arterioles is greater than that across capillaries.
    (e) **False** It is smaller, so blood flow velocity is higher in arterioles.

82 (a) **True** They conduct at around 4 m/s.
    (b) **True** This facilitates rapid conduction.
    (c) **False** Purkinje fibres travel to the apex before proceeding to the base of the heart.
    (d) **True** They spread depolarization rapidly over the entire ventricular myocardium.
    (e) **True** Damage to the cells (as in bundle-branch block) changes the pattern of spread of ventricular depolarization, and hence the shape of the QRS complex.

83 In the brachial artery:
    (a) Pulse waves travel at the same velocity as blood
    (b) Pulse pressure falls with decreasing elasticity of the wall
    (c) Pressure rises markedly when the artery is occluded distally
    (d) Pressure falls when the arm is raised above head level
    (e) Pulse pressures have a smaller amplitude than aortic pulse pressures

84 The tendency for blood flow to be turbulent increases when there is a decrease in blood:
    (a) Vessel diameter
    (b) Density
    (c) Flow velocity
    (d) Viscosity
    (e) Haemoglobin

85 Arterioles offer more resistance to flow than other vessels since they have:
    (a) Thicker muscular walls
    (b) Richer sympathetic innervation
    (c) Smaller internal diameters
    (d) A smaller total cross-sectional area
    (e) A greater pressure drop along their length

86 In the denervated heart, left ventricular stroke work increases when:
    (a) The end-diastolic length of the ventricular fibres increases
    (b) Peripheral resistance rises
    (c) Blood volume falls
    (d) Right ventricular output increases
    (e) The veins constrict

87 With increasing distance from the heart, arterial:
    (a) Walls contain more smooth muscle than elastic tissue
    (b) Flow has a greater tendency to be turbulent
    (c) Mean pressure tends to rise slightly
    (d) Pulse pressure tends to increase slightly
    (e) P_{O_2} falls appreciably

83 (a) **False** Pulse waves travel at about ten times the blood velocity.
    (b) **False** It rises; arterial elasticity normally damps the pulse pressure.
    (c) **False** Blood flows off rapidly via collaterals so that little pressure change occurs.
    (d) **True**   By the hydrostatic equivalent of the column of blood between it and the heart.
    (e) **False** Brachial arterial pulse pressures are greater due to the superimposition of waves reflected from the end of the arterial tree.

84 (a) **False** It is directly proportional to vessel diameter.
    (b) **False** It is directly proportional to fluid density.
    (c) **False** It is directly proportional to velocity.
    (d) **True**   It is inversely proportional to viscosity.
    (e) **True**   In anaemia, the increase in velocity and decrease in viscosity of blood in the hyperdynamic circulation promote turbulence; bruits may be heard over peripheral arteries.

85 (a) **False** Wall thickness is not a factor determining resistance.
    (b) **False** This suggests that smooth muscle tone may be varied reflexly.
    (c) **False** Capillaries have even smaller internal diameters.
    (d) **False** The aorta has a much smaller cross-sectional area.
    (e) **True**   The pressure drop across the arteriolar bed is larger than in other beds, indicating that arterioles are responsible for most of the circulation's vascular resistance.

86 (a) **True**   As stated in the Frank–Starling law.
    (b) **True**   This impedes ventricular outflow, thus increasing end-diastolic fibre length.
    (c) **False** This reduces cardiac filling pressure and hence end-diastolic fibre length.
    (d) **True**   This tends to increase left ventricular filling pressure and end-diastolic fibre length.
    (e) **True**   This also raises cardiac filling pressure.

87 (a) **True**   The relative amount of smooth muscle increases but that of elastic tissue falls.
    (b) **False** With the decreasing vessel diameter and flow velocity, the tendency decreases.
    (c) **False** It falls slightly; blood will only flow down a pressure gradient.
    (d) **True**   Distal arterial pulse pressure is increased by the superimposition of waves reflected back from the end of the arterial tree.
    (e) **False** Blood cannot release its oxygen until it reaches the exchange vessels.

88   In the estimation of cardiac output by an indicator dilution technique, the:
   (a)   Indicator must mix evenly with the entire blood volume
   (b)   Primary dilution curve may be followed by a secondary rise in indicator concentration
   (c)   Duration of the dilution curve shortens as cardiac output rises
   (d)   Mean indicator concentration under the curve increases as cardiac output rises
   (e)   Injection and monitoring devices may be placed in the pulmonary artery

89   In the estimation of cardiac output using the Fick principle:
   (a)   Pulmonary blood flow is measured
   (b)   The P_{O_2} of arterial and mixed venous blood are measured
   (c)   Oxygen uptake is estimated from alveolar P_{O_2} measurements
   (d)   Pulmonary arterial blood is sampled to measure the oxygen in mixed venous blood
   (e)   Pulmonary venous blood is sampled to measure the oxygen in arterial blood

90   Intravenous infusions of adrenaline and noradrenaline have a similar effect on:
   (a)   Skeletal muscle blood flow
   (b)   Renal blood flow
   (c)   Skin blood flow
   (d)   Diastolic arterial pressure
   (e)   Heart rate

91   When the heart suddenly stops beating (cardiac asystole):
   (a)   The physical signs are similar to those of ventricular fibrillation
   (b)   Consciousness is lost after 1–2 min
   (c)   Cardiac compression should be applied at a rate of 10–15/min
   (d)   Artificial ventilation of the lungs should be given
   (e)   Electric shocks across the thorax should be applied

92   Vasovagal fainting or syncope:
   (a)   Causes loss of consciousness
   (b)   Is associated with tachycardia
   (c)   Is associated with skeletal muscle vasodilation
   (d)   Is more likely to occur when standing than when lying down
   (e)   Is more likely to occur in a cold than in a hot environment

88 (a) **False** Complete mixing with the blood is required for estimation of blood volume.
    (b) **True**   The secondary peak is due to recirculation of indicator.
    (c) **True**   Due to the more rapid passage of indicator past the sampling site.
    (d) **False** It falls as the indicator is diluted in a bigger volume.
    (e) **True**   Pulmonary artery blood flow equals cardiac output.

89 (a) **True**   Pulmonary blood flow = right ventricular output = cardiac output.
    (b) **False** The O_2 contents of arterial and mixed venous blood are measured.
    (c) **False** To measure O_2 consumption, the subject breathes from a spirometer filled with oxygen with a CO_2 absorber in the circuit; an open circuit method may also be used.
    (d) **True**   The pulmonary artery is relatively easy to catheterize and the venous blood it contains is thoroughly mixed.
    (e) **False** Pulmonary veins are difficult to catheterize. Blood from any artery may be used since the O_2 content of peripheral arterial blood is the same as that in the pulmonary vein.

90 (a) **False** Adrenaline increases, and noradrenaline reduces skeletal muscle blood flow.
    (b) **True**   Both decrease renal blood flow.
    (c) **True**   Both cause cutaneous vasoconstriction.
    (d) **False** Noradrenaline raises diastolic pressure; adrenaline lowers it.
    (e) **False** Adrenaline increases heart rate; noradrenaline raises mean arterial pressure and causes reflex cardiac slowing.

91 (a) **True**   In neither case is there a useful cardiac output.
    (b) **False** Consciousness is lost within about 5 s.
    (c) **False** 60-80 compressions/min are needed to maintain flow to the brain.
    (d) **True**   This is needed to maintain oxygenation of the brain.
    (e) **False** This treatment (applied with a defibrillator) is given for ventricular fibrillation.

92 (a) **True**   Due to cerebral ischaemia caused by the abrupt fall in arterial pressure.
    (b) **False** Increased vagal activity slows the heart and reduces cardiac output.
    (c) **True**   Sympathetic vasodilator nerve activity thus reduces peripheral resistance.
    (d) **True**   Because of gravity, pressure in cerebral arteries is lower when standing than when lying down.
    (e) **False** The skin vasoconstriction in cold environments raises peripheral resistance.

93 Systemic hypertension may be caused by:
  (a) Hypoxia due to chronic respiratory failure
  (b) Excessive secretion of aldosterone
  (c) Excessive secretion of adrenocorticotrophic hormone
  (d) Myocardial thickening (hypertrophy) of the left ventricle
  (e) The rapid cardiac action of ventricular tachycardia

94 Peripheral differs from central circulatory failure in that:
  (a) Hypovolaemia is unusual
  (b) It leads to underperfusion of the tissues
  (c) Cardiac output is usually normal
  (d) Central venous pressure is low
  (e) Ventricular function is usually normal

95 In atrial fibrillation:
  (a) The electrocardiogram shows no evidence of atrial activity
  (b) Ventricular rate is lower than atrial rate
  (c) Respiratory sinus arrhythmia can usually be seen
  (d) The ventricular rate is irregular
  (e) The QRS complexes have an abnormal configuration

96 Severe systemic hypertension may result in:
  (a) Proliferation of myocardial cells in the left ventricle
  (b) Increased QRS voltage in certain leads
  (c) Increased coronary blood flow
  (d) Pulmonary oedema
  (e) Impaired vision

97 Auscultation of the heart can provide evidence of:
  (a) The direction of turbulent flow causing a murmur
  (b) Aortic stenosis, if there is a loud presystolic murmur in the aortic valve
        area
  (c) Mitral incompetence, if a systolic murmur is heard in the axilla
  (d) Ventricular septal defect, if a loud diastolic murmur is heard
  (e) Mitral stenosis, if early diastolic and presystolic murmurs are heard

93 **(a) False**   This constricts blood vessels in the lungs causing pulmonary hypertension but dilates systemic vessels.

    (b) True   Salt and water retention by the kidneys expands ECF and hence blood volume and cardiac output.

    (c) True   The resulting secretion of cortisol also causes salt and water retention.

    (d) False   This is a consequence of hypertension, not a cause.

    (e) False   This ineffective pumping in ventricular fibrillation causes severe hypotension.

94 **(a) False**   Hypovolaemia due to severe haemorrhage is a common cause of peripheral circulatory failure; blood volume may be normal in central circulatory failure.

    (b) False   Both types of failure lead to underperfusion of the tissues.

    (c) False   It is usually reduced in both types of failure.

    (d) True   It is usually raised in central circulatory failure.

    (e) True   Reduced ventricular function is the cause of central circulatory failure.

95 **(a) False**   Small rapid waves indicate the atrial fibrillation.

    (b) True   Some impulses are filtered out by the atrioventricular node.

    (c) False   Sinus arrhythmia indicates normal sinus rhythm.

    (d) True   Due to the irregularity of the impulses passing through the AV node.

    (e) False   QRS complexes are normal since the pattern of ventricular depolarization is normal.

96 **(a) False**   The cells increase in size (hypertrophy), not in number (hyperplasia).

    (b) True   Due to the left ventricular hypertrophy.

    (c) True   Due to increased left ventricular work.

    (d) True   Due to left ventricular failure.

    (e) True   Due to damage to retinal blood vessels.

97 **(a) True**   The direction in which the murmur is conducted indicates the direction of flow.

    (b) False   The characteristic murmur is a systolic murmur conducted to the neck vessels.

    (c) True   This is the direction of flow of the regurgitant blood.

    (d) False   The murmur occurs during ventricular contraction and is therefore systolic.

    (e) True   Mitral flow is greatest in early diastole but rises again during atrial systole.

98 The electrocardiogram shows:
  (a) Irregular P waves in atrial flutter
  (b) Regular QRS complexes in atrial fibrillation
  (c) Regular QRS complexes in complete heart block
  (d) High-voltage R waves over the right ventricle in right ventricular hypertrophy
  (e) An irregular saw-tooth appearance in ventricular fibrillation

99 The jugular venous:
  (a) Pulse is not visible in normal healthy people
  (b) Pulse has greater amplitude in patients with tricuspid incompetence
  (c) Pulse can vary widely in amplitude in patients with complete heart block
  (d) Pressure is raised in patients with right ventricular failure
  (e) Pressure is commonly raised in patients with mediastinal tumours

100 In heart failure:
  (a) The resting cardiac output may be higher than normal
  (b) The arteriovenous oxygen difference during exercise is less than in normal people
  (c) There is sodium retention
  (d) Oedema occurs in dependent parts of the body
  (e) Pulmonary oedema occurs when pulmonary capillary pressure doubles

101 Respiratory failure (low arterial P_{O_2}; raised arterial P_{CO_2}) leads to:
  (a) Raised pulmonary artery pressure (pulmonary hypertension)
  (b) Right ventricular failure
  (c) Low-voltage P waves in the electrocardiogram
  (d) Decreased cerebral blood flow
  (e) Warm hands and feet

102 Pain due to poor coronary blood flow (angina) may be relieved by:
  (a) Cutting the sympathetic nerve trunks supplying the heart
  (b) Correcting anaemia if present
  (c) Providing the patient with a cold environment
  (d) β-Adrenoceptor stimulating drugs
  (e) Drugs causing peripheral vasodilation

98 (a) **False** In atrial flutter, P waves have a high but regular frequency (about 300/min).

(b) **False** Ventricular beats are irregular in rate and strength since impulses pass through the AV node in a random fashion.

(c) **True** The beats generated by ventricular pacemakers have slow but regular frequency.

(d) **True** The increased muscle bulk generates bigger voltages during depolarization.

(e) **True** Due to chaotic electrical activity in the ventricles.

99 (a) **False** It can be seen in healthy people when they lie almost flat.

(b) **True** Systolic regurgitation of right ventricular blood can cause giant waves.

(c) **True** Periodic giant waves are seen when atrial and ventricular systoles coincide.

(d) **True** This is an important sign in right ventricular failure; venous pulsation is present.

(e) **True** The jugular veins are distended but there is little or no pulsation.

100 (a) **True** In high output failures; however, the ability to raise cardiac output in exercise is impaired in all types of failure.

(b) **False** With inadequate output, desaturation of blood in the tissues increases.

(c) **True** This increases extracellular fluid and hence blood volume.

(d) **True** The back pressure in veins raises capillary hydrostatic pressure and results in oedema in dependent parts where venous pressure is already raised due to gravity.

(e) **False** When pulmonary capillary pressure (about 5 mmHg) doubles, it is still less than plasma oncotic pressure (25 mmHg) so fluid does not accumulate in the alveoli.

101 (a) **True** Hypoxia causes generalized pulmonary vasoconstriction.

(b) **True** Pulmonary hypertension can lead to right ventricular failure ('cor pulmonale').

(c) **False** Atrial hypertrophy in cor pulmonale results in prominent P waves.

(d) **False** CO_2 dilates cerebral blood vessels.

(e) **True** CO_2 is a vasodilator and 'cor pulmonale' is a 'high-output' failure.

102 (a) **True** Pain sensory fibres from the heart travel with the sympathetic nerves.

(b) **True** In anaemia, the capacity of the blood to deliver oxygen is decreased but cardiac work increases due to the rise in cardiac output.

(c) **False** Cold vasoconstriction raises arterial pressure and so increases myocardial work.

(d) **False** These increase heart rate and force and so increase myocardial work.

(e) **True** By reducing arterial pressure, vasodilator drugs such as nitrates reduce myocardial work.

103 Narrowing of the lumen of major arteries supplying the leg is associated with:
   (a) Pain in the calf during exercise which is relieved by rest
   (b) Growth of collateral vessels
   (c) Reduction in the duration of reactive hyperaemias in the calf
   (d) Delayed healing of skin damage in the leg
   (e) Reduced arterial pulse amplitude at the ankle

104 Arterial pulse contours which have:
   (a) Sharp peaks indicate rapid left ventricular ejection
   (b) Greatly increased pulse pressures are seen in patients with mitral incompetence
   (c) Slowly rising systolic phases are seen in patients with aortic stenosis
   (d) Varying beat-to-beat amplitude are seen in patients with atrial fibrillation
   (e) Rapid run-offs and low diastolic pressure suggest high peripheral resistance

105 Murmurs (or bruits) may be detected by auscultation over:
   (a) Vessels in which there is turbulence
   (b) Large arteries in healthy adults
   (c) Dilations (aneurysms) in arteries
   (d) Constrictions (stenoses) in arteries
   (e) The heart in healthy young adults in early diastole

106 Factors ensuring that ventricular muscle has an adequate oxygen supply include the:
   (a) Good functional anastomoses that exist between adjacent coronary arteries
   (b) Structural arrangements which prevent vascular compression during systole
   (c) High oxygen extraction rate from blood circulating through the myocardium
   (d) Sympathetic vasodilator nerve supply to ventricular muscle
   (e) Fall in coronary vascular resistance during exercise

107 When the AV bundle is completely interrupted, as in complete heart block, the:
   (a) Atrial beat becomes irregular
   (b) PR interval shows beat-to-beat variability
   (c) Ventricular filling shows beat-to-beat variability
   (d) QRS complex shows beat-to-beat variability
   (e) Ventricular rate falls below 50 beats/min

103 (a) True This is intermittent claudication; during exercise in ischaemic limbs, pain metabolites accumulate in muscle and stimulate local pain receptors.

(b) True When the major arteries are obstructed, collateral vessels open up to help maintain blood flow to the ischaemic tissues.

(c) False Though the hyperaemias have smaller peak values, their durations are longer.

(d) True Cuts and ulcers are slow to heal because the supply of nutrients is impaired.

(e) True Pulses may be absent with severe narrowing.

104 (a) True As seen in strenuous exercise.

(b) False This is typical of aortic incompetence.

(c) True Due to slow expulsion of blood from the ventricle past the stenosed valve.

(d) True The variable filling of the ventricle results in variable stroke output.

(e) False It suggests a low peripheral resistance.

105 (a) True Turbulent vibrations generate sound waves.

(b) False Flow is laminar rather than turbulent in normal large arteries.

(c) True Turbulence is set up as blood flows into the dilated segment.

(d) True Turbulence is set up as blood squirts through the constricted segment.

(e) True Rapid ventricular filling in early diastole can cause turbulence in healthy young adults and generate a sound (the third heart sound).

106 (a) False Coronary arteries are functional 'end arteries' and have few anastomotic connections; sudden occlusion of an artery usually leads to local muscle death.

(b) False Coronary vessels are compressed by the contracting myocardium in systole.

(c) True The extraction rate is about 75%.

(d) False Reflex vasodilation is not important in regulating coronary blood flow.

(e) True The rise in metabolic activity in the exercising heart provides the vasodilator metabolites which adapt coronary flow to supply myocardial oxygen needs.

107 (a) False The atria continue to beat regularly at normal sinus rate.

(b) True In complete block, atria and ventricles beat independently at different rates.

(c) True Due to loss of the normal sequence of the atrial and ventricular contractions.

(d) False There is usually a single ventricular pacemaker giving an abnormal but regular QRS complex.

(e) True The ventricular intrinsic rate of beating is 30–40/min.

108 Aortic valve incompetence may cause:
   (a) Increase in arterial pulse pressure
   (b) Systolic murmurs in the aortic valve area
   (c) Hypertrophy of left ventricular muscle
   (d) Decreased myocardial blood flow
   (e) Left ventricular failure

109 Ventricular extrasystoles:
   (a) Are usually associated with a normal QRS complex
   (b) From the same focus have similar QRS complexes
   (c) Usually occur following a compensatory pause
   (d) May fail to produce a pulse at the wrist
   (e) Indicate serious heart disease

110 Pulmonary embolism (blood clots impacting in lung blood vessels) usually decreases:
   (a) Pulmonary vascular resistance
   (b) Left atrial pressure
   (c) Right atrial pressure
   (d) Ventilation:perfusion ratios in the affected lung
   (e) P_{O_2} in pulmonary venous blood

111 Hardening of the arterial walls tends to raise:
   (a) Arterial compliance
   (b) Systolic arterial pressure
   (c) Diastolic arterial pressure
   (d) Peripheral resistance
   (e) Arterial pulse wave velocity

112 The 'A' wave of venous pulsation in the neck is:
   (a) Caused by atrial systole
   (b) Seen just after the carotid artery pulse
   (c) Exaggerated in atrial fibrillation
   (d) Exaggerated in tricuspid stenosis
   (e) Exaggerated in complete heart block

108 **(a) True**   Diastolic pressure is abnormally low due to regurgitation of aortic blood into the left ventricle in diastole.
(b) False   Blood regurgitating in diastole causes diastolic turbulence.
(c) True   The greater stroke volume needed to compensate for regurgitating blood increases ventricular work-load.
(d) False   Flow increases as ventricular work increases.
(e) True   A persistent increase in ventricular work-load can lead to ventricular failure.

109 **(a) False**   Extrasystoles are associated with a prolonged abnormal QRS complex; the impulse pathway from the ectopic focus over the myocardium is abnormal and slow.
(b) True   If the focus is the same, the pathway for ventricular excitation will be the same.
(c) False   They are followed by a compensatory pause; the next normal beat may reach the ventricles when they are refractory – a beat is lost.
(d) True   If they occur early in diastole, poor ventricular filling results in weak contractions and small pulses.
(e) False   They occur occasionally in many normal hearts.

110 **(a) False**   Vascular obstruction tends to increase pulmonary vascular resistance and cause pulmonary hypertension.
(b) True   Due to the fall in pulmonary blood flow, atrial pressure tends to fall.
(c) False   The obstruction tends to dam back blood in the right heart.
(d) False   It rises since pulmonary capillary perfusion falls.
(e) False   Blood that traverses pulmonary capillaries is adequately oxygenated; cyanosis in patients with pulmonary embolism is usually peripheral cyanosis due to low cardiac output.

111 **(a) False**   Compliance, the change of arterial volume per unit pressure change, decreases.
(b) True   Systolic ejection causes greater pressure rise when arteries are less distensible.
(c) False   Poor elastic recoil in diastole allows diastolic pressure to fall further.
(d) False   Stiffness of the wall is not a factor determining vascular resistance.
(e) True   Vibration travels faster in stiff than in lax structures.

112 **(a) True**   The pressure wave due to atrial contraction passes up freely into the neck.
(b) False   Atrial systole precedes the ventricular systole that generates the carotid pulse.
(c) False   It is absent – there is no effective atrial systole in atrial fibrillation.
(d) True   Right atrial contraction is more forceful to overcome valvular resistance.
(e) True   If the atrial and ventricular systoles coincide, the A and C waves merge to give a giant wave.

113 Left ventricular failure tends to cause an increase in:
   (a) Left atrial pressure
   (b) Left ventricular ejection fraction
   (c) Pulmonary capillary pressure
   (d) Lung compliance
   (e) Pulmonary oedema when the patient stands up

114 In otherwise healthy people, local tissue death follows obstruction of:
   (a) An internal carotid artery
   (b) A renal artery
   (c) A femoral artery
   (d) A retinal artery
   (e) The hepatic portal vein

115 In the measurement of forearm blood flow by venous occlusion plethysmography:
   (a) A continuous record is made of forearm volume
   (b) A collecting cuff is applied to the wrist
   (c) It is assumed that venous outflow is arrested by the collecting cuff during measurement
   (d) The collecting cuff pressure should be greater than diastolic but less than systolic arterial pressure
   (e) The forearm must be kept below heart level during the measurements

113 (a) True   Due to inadequate emptying of the left ventricle in systole.
(b) False Stroke volume falls; end diastolic volume rises.
(c) True   The rise may cause pulmonary oedema if pressure exceeds the colloid osmotic pressure of the plasma proteins.
(d) False It falls; congestion of pulmonary vessels with blood makes the lungs stiffer.
(e) False The decrease in venous return on standing up may relieve pulmonary congestion and hence dyspnoea.

114 (a) False Flow through the circle of Willis normally maintains the viability of the tissue.
(b) True   There is no significant collateral circulation.
(c) False There is adequate collateral circulation; in the presence of advanced arterial disease, sudden obstruction may cause gangrene of the tissues.
(d) True   The collateral circulation is not good enough to prevent this.
(e) False The liver has a dual blood supply; hepatic artery flow can maintain viability.

115 (a) True   Plethysmography means volume measurement.
(b) False It is applied to the upper arm.
(c) True   When this is the case, the volume increase equals arterial inflow.
(d) False It must be below diastolic so as not to interfere with arterial inflow.
(e) False Below heart level, the veins fill with blood and are unable to accommodate more during venous occlusion; the forearm must be above heart level.

3 RESPIRATORY SYSTEM

116 In a person breathing normally at rest with an environmental temperature of 25°C, the partial pressure of:
  (a) CO_2 in alveolar air is about twice that in room air
  (b) Water vapour pressure in alveolar air is less than half the alveolar P_{CO_2} level
  (c) Water vapour pressure in alveolar air is greater than that in room air even at 100% humidity
  (d) O_2 in expired air is greater than in alveolar air
  (e) CO_2 in mixed venous blood is greater than in alveolar air

117 As blood passes through systemic capillaries:
  (a) pH rises
  (b) HCO_3^- ions pass from red cells to plasma
  (c) Cl^- ion concentration in red cells falls
  (d) Its oxygen dissociation curve shifts to the right
  (e) Its ability to deliver oxygen to the tissues is enhanced

118 The respiratory centre:
  (a) Is in the hypothalamus
  (b) Sends impulses to inspiratory muscles during quiet breathing
  (c) Sends impulses to expiratory muscles during quiet breathing
  (d) Is involved in the swallowing reflex
  (e) Is not involved in the vomiting reflex

119 The carotid bodies:
  (a) Are stretch receptors in the walls of the internal carotid arteries
  (b) Have a blood flow per unit volume similar to that in the brain
  (c) Are influenced more by blood P_{O_2} than by its oxygen content
  (d) Generate more afferent impulses when blood H^+ ion concentration rises
  (e) And the aortic bodies are mainly responsible for the increased ventilation in hypoxia

120 Pulmonary surfactant increases:
  (a) The surface tension of the fluid lining alveolar walls
  (b) Lung compliance
  (c) In effectiveness as the lungs are inflated
  (d) In amount when pulmonary blood flow is interrupted
  (e) In amount in fetal lungs during the last month of pregnancy

121 As people age, there is usually a decrease in their:
  (a) Ratio of lung residual volume to vital capacity
  (b) Percentage of vital capacity expelled in 1 s
  (c) Lung volume level at which small airways start to close during expiration
  (d) Lung elasticity
  (e) Resting arterial blood P_{O_2}

116 (a) **False** Room air P_{CO_2} (0.2 mmHg; 0.03 kPa) is negligible compared with alveolar air P_{CO_2} (40 mmHg; 5.3 kPa).

(b) **False** Alveolar H_2O vapour pressure at 37°C is 47 mmHg (6.3 kPa).

(c) **True** Alveolar air is saturated with water vapour; at 37°C it exceeds that at 25°C.

(d) **True** Expired air is alveolar air plus dead space air.

(e) **True** This is necessary for CO_2 excretion by diffusion.

117 (a) **False** It falls, due largely to CO_2 uptake.

(b) **True** Incoming CO_2 is converted to HCO_3^- (carbonic anhydrase in RBCs catalyses the reaction $CO_2 + H_2O \rightleftharpoons H^+ + HCO_3^-$; HCO_3^- migrates out as its concentration rises.

(c) **False** It rises as Cl^- moves in to replace departing HCO_3^- in the 'chloride shift'.

(d) **True** Due largely to the rise in blood P_{CO_2}.

(e) **True** Due to the shift of the oxygen dissociation curve.

118 (a) **False** It is in the medulla oblongata.

(b) **True** Causing expansion of the thorax cage.

(c) **False** Expiration at rest is passive.

(d) **True** It is inhibited during swallowing, so preventing aspiration of food.

(e) **False** Expiratory movements, including abdominal muscle contraction with the oesophagus relaxed and the glottis closed help expel gastric contents in vomiting.

119 (a) **False** The stretch receptors in internal carotid arteries are carotid sinus baroreceptors.

(b) **False** They have the greatest flow rate/unit volume yet described in the body.

(c) **True** They are not excited in anaemia where P_{O_2} is normal but O_2 content is low.

(d) **True** Acidosis stimulates ventilation.

(e) **True** When carotid and aortic bodies are denervated, hypoxia depresses respiration.

120 (a) **False** It decreases surface tension.

(b) **True** It permits the lungs to be more easily inflated.

(c) **False** The decreasing effect as lungs inflate helps prevent over-inflation.

(d) **False** It decreases; this may lead to local collapse of the lung.

(e) **True** Premature babies may have breathing problems due to surfactant deficiency.

121 (a) **False** Residual volume increases and vital capacity decreases.

(b) **True** This gradually falls with normal ageing.

(c) **False** The 'closing volume' increases with age.

(d) **True** This increases residual and closing volumes.

(e) **True** There is, however, little change in oxygen saturation.

122 During inspiration:
   (a) Intrapleural pressure is lowest at mid-inspiration
   (b) Intrapulmonary pressure is lowest at end-expiration
   (c) Intraoesophageal pressure is lowest at mid-inspiration
   (d) The rate of air flow is greatest at end-inspiration
   (e) The lung volume/intrapleural pressure relationship is the same as in expiration

123 Carbon dioxide:
   (a) Is carried as carboxyhaemoglobin on the haemoglobin molecule
   (b) Uptake by the blood increases its oxygen-binding power
   (c) Uptake by the blood leads to similar increases in H^+ and HCO_3^- ion concentrations
   (d) Stimulates ventilation when breathed at a concentration of 20%
   (e) Content is greater than oxygen content in arterial blood

124 In normal lungs:
   (a) The rate of alveolar ventilation at rest exceeds the rate of alveolar capillary perfusion
   (b) The ventilation/perfusion (V/P) ratio exceeds 2.0 during maximal exercise
   (c) The V/P ratio is higher at the apex than at the base of the lungs when a person is standing
   (d) Oxygen transfer can be explained by passive diffusion
   (e) Dead space increases during inspiration

125 Alveolar ventilation is increased by breathing:
   (a) 21% O_2 and 79% N_2
   (b) 17% O_2 and 83% N_2
   (c) 2% CO_2 and 98% O_2
   (d) 10% CO_2 and 90% O_2
   (e) A gas mixture which raises arterial P_{CO_2} by 10%

126 Bronchial smooth muscle contracts in response to:
   (a) Bronchial mucosal irritation
   (b) Local beta adrenoceptor stimulation
   (c) A fall in bronchial P_{CO_2}
   (d) Inhalation of cold air
   (e) Circulating noradrenaline

127 In early inspiration there is a fall in:
   (a) Intrapulmonary pressure
   (b) Intrathoracic pressure
   (c) Intra-abdominal pressure
   (d) Dead space P_{O_2}
   (e) Pressure in the superior vena cava

122 **(a) False** It is lowest at the end of inspiration.
    (b) False This is lowest around the middle of inspiration.
    (c) False Its pattern is similar to that of intrapleural pressure.
    (d) False It is greatest at mid-inspiration; it depends on the mouth:alveoli pressure gradient.
    (e) False Volume changes lag behind pressure changes to give a hysteresis loop.

123 **(a) False** It is carried as carbaminohaemoglobin; carboxyhaemoglobin is the combination of haemoglobin with carbon monoxide.
    (b) False It decreases as the oxygen dissociation curve is shifted to the right.
    (c) False HCO_3^- ions increase more; H^+ ions are largely buffered by haemoglobin.
    (d) False Breathing 20% CO_2 causes respiratory depression; stimulation of respiration by CO_2 is maximal when breathed at concentrations of around 5%.
    (e) True CO_2 content in arterial blood is about 500 ml/litre; O_2 content is about 200 ml/litre.

124 **(a) False** Alveolar ventilation at rest is about 4 litres/min; perfusion is about 5 litres/min.
    (b) True In maximal exercise, alveolar ventilation may rise to about 80 litres/min whereas alveolar perfusion (cardiac output) rises to about 25 litres/min.
    (c) True Perfusion decreases from base to apex; ventilation does also but to a lesser extent.
    (d) True There is no evidence of O_2 secretion in the lungs.
    (e) True The trachea and bronchi expand as the lungs expand.

125 **(a) False** This is the normal composition of air.
    (b) False The O_2 level must fall to around 15% before breathing is stimulated.
    (c) True The stimulating effect of high P_{CO_2} is little affected by high P_{O_2} levels.
    (d) False This level of carbon dioxide depresses breathing.
    (e) True This is enough to double the volume breathed per minute.

126 **(a) True** Via a reflex with a parasympathetic efferent pathway.
    (b) False This leads to relaxation.
    (c) True This tends to limit local overventilation.
    (d) True This is minimized normally by the warming of air in the upper airways.
    (e) False This causes relaxation via beta receptors.

127 **(a) True** This creates a pressure gradient between mouth and lungs.
    (b) True Due to an increase in the dimensions of the thoracic cage.
    (c) False It rises due to descent of the diaphragm.
    (d) False It rises as inspired air replaces alveolar air.
    (e) True It falls as intrathoracic pressure falls.

128 Compliance of the lungs is greater:
  (a) When they are expanded above their normal tidal volume range
  (b) In adults than in infants
  (c) Than the compliance of the lungs and thorax together
  (d) When they are filled with normal saline than when they are filled with air
  (e) In standing than in recumbent subjects

129 At a high altitude where atmospheric pressure is halved, there is an increase in:
  (a) Pulmonary ventilation
  (b) Alveolar H_2O vapour pressure
  (c) Arterial P_{O_2}
  (d) Arterial pH
  (e) Cerebral blood flow

130 During inspiration:
  (a) Venous return to the heart is increased
  (b) More energy is expended than during expiration
  (c) Lung expansion is assisted by surface tension forces in the alveoli
  (d) Lung expansion begins when intrapleural pressure falls below atmospheric
  (e) Lung expansion ends when intrapulmonary pressure falls to atmospheric

131 The residual volume is:
  (a) The gas remaining in the lungs at the end of a full expiration
  (b) Greater on average in men than in women
  (c) 3–4 litres on average in young adults
  (d) Measured directly using a spirometer
  (e) Smaller in old than in young people

132 A rise in arterial P_{CO_2} leads to an increase in:
  (a) Ventilation due to stimulation of peripheral chemoreceptors
  (b) Ventilation due to stimulation of central chemoreceptors
  (c) Arterial pressure
  (d) Cerebral blood flow
  (e) The plasma bicarbonate level

133 Ventilation is increased during:
  (a) Periods when cerebrospinal fluid pH is reduced
  (b) Chronic renal failure
  (c) Periods when plasma bicarbonate level is raised
  (d) Deep sleep
  (e) Exercise because of the ensuing fall in arterial P_{O_2}

128 (a) **False** Compliance is maximal in the tidal volume range.
(b) **True** Compliance is extremely small in infants.
(c) **True** It is nearly twice as great.
(d) **True** There are no surface tension forces to overcome in fluid-filled lungs.
(e) **True** Lungs are less stiff when their blood content falls.

129 (a) **True** Due to stimulation of chemoreceptors by oxygen lack.
(b) **False** This remains at the saturated pressure at body temperature.
(c) **False** The fall in arterial P_{O_2} stimulates the carotid bodies to increase ventilation.
(d) **True** Hyperventilation causes respiratory alkalosis.
(e) **False** The fall in P_{CO_2} causes cerebral vasoconstriction.

130 (a) **True** By decreasing intrathoracic venous pressure.
(b) **True** Expiration is assisted by the elastic recoil of the lungs and thoracic cage.
(c) **False** Surface tension is a force to be overcome in inspiration.
(d) **False** Expansion begins when intrapulmonary pressure falls below atmospheric.
(e) **True** There is no pressure gradient at this point to drive air.

131 (a) **True** This is its definition.
(b) **True** Men, on average, have bigger thoracic cages than women.
(c) **False** It is around 1–1.5 litres
(d) **False** Residual air cannot be exhaled; it is measured indirectly by a dilution technique.
(e) **False** It increases with age since the elastic recoil of the lungs decreases with age.

132 (a) **True** Via the carotid and aortic bodies.
(b) **True** The central effect predominates.
(c) **True** Reflex vasoconstriction and cardiac stimulation predominate over the direct vasodilator effects of CO_2.
(d) **True** The cerebral vessels are little affected by sympathetic reflexes so the direct vasodilator effect of CO_2 predominates.
(e) **True** To compensate for the respiratory acidosis.

133 (a) **True** CO_2 acts centrally by reducing the pH of CSF.
(b) **True** Because of the ensuing metabolic acidosis.
(c) **False** The metabolic alkalosis depresses ventilation.
(d) **False** Ventilation falls during a deep sleep.
(e) **False** Arterial P_{O_2} is well maintained in exercise.

134 Voluntary hyperventilation increases the:
   (a) Negative charge on the plasma proteins
   (b) Level of ionized calcium in blood
   (c) Alveolar P_{O_2} three-fold when ventilation is increased three-fold
   (d) Arterial blood oxygen saturation by 10–15% when ventilation is increased by 10–15%
   (e) Renal excretion of bicarbonate

135 If the carotid and aortic chemoreceptors are denervated:
   (a) Increasing alveolar P_{CO_2} by 25% fails to stimulate ventilation
   (b) Halving the alveolar P_{O_2} fails to stimulate ventilation
   (c) The resting ventilation rate is depressed by more than 40%
   (d) Ventilation does not increase during exercise
   (e) The ability to adapt to life at high altitude is impaired

136 Pulmonary:
   (a) Arterial mean pressure is about one-sixth systemic mean arterial pressure
   (b) Blood flow/min is similar to systemic blood flow/min
   (c) Vascular resistance is about 50% that of systemic vascular resistance
   (d) Vascular capacity is similar to systemic vascular capacity
   (e) Arterial pressure increases by about 50% when cardiac output rises by 50%

137 Carbon dioxide is carried in the blood in:
   (a) Combination with the haemoglobin molecule
   (b) Combination with plasma proteins
   (c) Physical solution in plasma
   (d) Greater quantity in red blood cells than in plasma
   (e) Greater quantity as HCO_3^- ions than as other forms

138 A shift of the oxygen dissociation curve of blood to the right:
   (a) Occurs in the pulmonary capillaries
   (b) Occurs if blood temperature rises
   (c) Favours oxygen delivery to the tissues
   (d) Favours oxygen uptake from the lungs by alveolar capillary blood
   (e) Increases the P_{50} (the P_{O_2} value giving 50% blood oxygen saturation)

139 The work of breathing increases when:
   (a) Lung compliance increases
   (b) The subject exercises
   (c) The rate of breathing increases even though the minute volume stays constant
   (d) The subject lies down
   (e) Functional residual capacity increases

134 (a) True It makes it more negative by raising blood pH.
(b) False It decreases as protein binding of calcium increases.
(c) False Alveolar P_{O_2} cannot exceed the P_{O_2} of the air being breathed (150 mmHg).
(d) False The blood is normally almost fully saturated (95–98%).
(e) True To compensate for the respiratory alkalosis.

135 (a) False Raised P_{CO_2} can still stimulate breathing by acting on central chemoreceptors.
(b) True In fact, hypoxia depresses ventilation by its action on the respiratory centre.
(c) False Normal ventilation is driven mainly by the effect of CO_2 on central chemoreceptors.
(d) False Central and other peripheral effects increase respiratory drive in exercise.
(e) True The individual is unable to increase ventilation in response to hypoxia.

136 (a) True About 15 compared with 90 mmHg in the systemic circuit.
(b) True Otherwise blood would accumulate in one or other bed.
(c) False From (a) and (b) above, it is only about 17% of systemic resistance.
(d) False Its capacity is about one-third of systemic capacity.
(e) False Pulmonary vascular resistance falls when cardiac output rises, due possibly to endothelial release of nitric oxide.

137 (a) True Attached to amino groups as carbaminohaemoglobin (Hb-NH.COOH).
(b) True Attached to amino groups as carbaminoprotein (Pr-NHCOOH).
(c) True It is the CO_2 in solution that exerts the P_{CO_2} and is diffusible.
(d) False Plasma carries the greater quantity, mainly as bicarbonate.
(e) True Bicarbonate accounts for 80–90% of the total CO_2 in blood.

138 (a) False It occurs in the systemic capillaries
(b) True This occurs when tissue metabolic activity increases.
(c) True More oxygen is released at any given P_{O_2}.
(d) False Less oxygen is taken up at any given P_{O_2}.
(e) True For example, a rise in temperature raises the P_{50}.

139 (a) False Compliant lungs are easier to inflate.
(b) True Since the rate and depth of ventilation are increased.
(c) True The work of breathing is minimal around the normal rate of breathing.
(d) True The increase in pulmonary blood volume increases lung stiffness and abdominal pressure on the diaphragm rises.
(e) True Lung compliance falls at higher lung volumes.

140 The compliance of the lungs and chest wall is:
  (a) Expressed as volume change per unit change in pressure
  (b) Minimal during quiet breathing
  (c) Increased by the surface tension of the fluid lining the alveoli
  (d) Increased by surfactant
  (e) Changed by parallel displacement of the line relating lung volume to distending pressure

141 Respiratory dead space:
  (a) Saturates inspired air with water vapour before it reaches the alveoli
  (b) Removes all particles from inspired air before it reaches the alveoli
  (c) Decreases when blood catecholamine levels rise
  (d) Decreases during a deep inspiration
  (e) Decreases during a cough

142 Vital capacity is:
  (a) The volume of air expired from full inspiration to full expiration
  (b) Reduced as one grows older
  (c) Greater in men than in women of the same age and height
  (d) Related more to total body mass than to lean body mass
  (e) The sum of the inspiratory and expiratory reserve volumes

143 In pulmonary capillary blood:
  (a) Carbonic anhydrase in erythrocytes catalyses the formation of H^+ and HCO_3^-
  (b) Hydrogen ions dissociate from haemoglobin
  (c) The rise in P_{O_2} is of greater magnitude than the fall in P_{CO_2}
  (d) The oxygen content is linearly related to alveolar P_{O_2}
  (e) The pH is lower than in blood in the pulmonary artery

144 Oxygen debt is:
  (a) The amount of O_2 consumed after cessation of exercise
  (b) Incurred because the pulmonary capillary walls limit O_2 uptake during exercise
  (c) Possible since skeletal muscle can function temporarily without oxygen
  (d) Associated with a rise in blood lactate
  (e) Associated with metabolic acidosis

145 The CO_2 dissociation curve for whole blood shows that:
  (a) Its shape is sigmoid
  (b) Blood saturates with CO_2 when P_{CO_2} exceeds normal alveolar levels
  (c) Blood contains some CO_2 even when the P_{CO_2} is zero
  (d) Oxygenation of the blood drives CO_2 out of the blood
  (e) Adding CO_2 to the blood drives O_2 out of the blood

140 (a) **True** The normal value is about 0.1 litre/cm H_2O (1 litre/kPa).
    (b) **False** It is maximal over this range of chest movement.
    (c) **False** The surface tension of alveolar fluid decreases compliance.
    (d) **True** This decreases the surface tension.
    (e) **False** The slope of this line indicates compliance and is unchanged by a parallel shift.

141 (a) **True** This prevents drying of the alveolar surface.
    (b) **False** Particles less than about 2 μm can reach the alveoli
    (c) **False** Catecholamines relax airway muscle and constrict mucosal vessels.
    (d) **False** Expansion of the lungs expands airways as well as alveoli.
    (e) **True** This allows air to be expelled at high airway velocity.

142 (a) **True** This is how it is usually measured.
    (b) **True** It falls on average by about 1 litre between age 20 and age 70.
    (c) **True** By 0.5–1 litre.
    (d) **False** It is related closely to lean body mass.
    (e) **False** The tidal volume must be added.

143 (a) **False** It catalyses the conversion of H_2CO_3 to CO_2 and H_2O.
    (b) **True** Oxygenation of haemoglobin favours this dissociation.
    (c) **True** P_{O_2} rises by at least 40 mmHg whereas P_{CO_2} falls by only 6 mmHg.
    (d) **False** It follows the sigmoid oxygen dissociation curve.
    (e) **False** It is higher due to the diffusion of CO_2 out of the blood to the alveoli.

144 (a) **False** It is the amount of O_2 in excess of resting needs consumed after exercise stops.
    (b) **False** O_2 uptake across pulmonary capillary walls is not usually diffusion limited.
    (c) **True** The O_2 debt incurred during anaerobic metabolism is repaid later.
    (d) **True** This is generated during anaerobic metabolism.
    (e) **True** Due to the fall in blood HCO_3^- used up buffering lactic acid.

145 (a) **False** It does not reach a plateau; it is the O_2 dissociation curve that is sigmoid in shape.
    (b) **False** Content continues to rise as P_{CO_2} rises above normal alveolar levels.
    (c) **False** CO_2 content is zero when P_{CO_2} is zero.
    (d) **True** Oxyhaemoglobin is a stronger acid than reduced haemoglobin; the liberated H^+ ions drive the reaction $H^+ + HCO_3^- \rightarrow H_2CO_3 \rightarrow CO_2 + H_2O$ to the right.
    (e) **False** This is shown by the oxygen dissociation curve.

146 The oxygen content of mixed venous blood is:
   (a) Measured using blood sampled from the right atrium
   (b) Increased during generalized muscular exercise
   (c) Increased in a warm environment
   (d) Increased in cyanide poisoning
   (e) Decreased in circulatory failure

147 Bronchial asthma is likely to be relieved by:
   (a) Stimulation of cholinergic receptors
   (b) Stimulation of β adrenoceptors
   (c) Histamine aerosols
   (d) Drugs which stabilize mast cell membranes
   (e) Glucocorticoids

148 Air in the pleural cavity (pneumothorax):
   (a) Allows intrapleural pressure to rise to atmospheric pressure
   (b) Causes the underlying lung to collapse by compressing it
   (c) Increases the functional residual capacity
   (d) Leads to a slight outward movement of the chest wall
   (e) Reduces vital capacity

149 A patient with chronic respiratory failure:
   (a) Shows an enhanced respiratory sensitivity to inhaled carbon dioxide
   (b) Shows little or no respiratory response to hypoxia
   (c) Is likely to have a low blood bicarbonate level
   (d) Responds well when given 100% oxygen to breathe
   (e) Must have been breathing oxygen-enriched air if alveolar P_{CO_2} is 150 mmHg (20 kPa)

150 Loss of pulmonary elastic tissue in 'emphysema' reduces:
   (a) Physiological dead space
   (b) Anatomical dead space
   (c) Residual volume
   (d) Vital capacity
   (e) The percentage of the vital capacity expired in 1 s

151 Complete obstruction of a major bronchus usually results in:
   (a) Collapse of the alveoli supplied by the bronchus
   (b) A rise in local intrapleural pressure
   (c) An increase in physiological dead space
   (d) An increase in blood flow to the lung tissue supplied by the bronchus
   (e) Cyanosis

146 (a) False For adequate mixing, pulmonary artery blood is needed.
 (b) False It falls due to increased oxygen extraction by the active muscles.
 (c) True Due to large volumes of well oxygenated blood returning from skin.
 (d) True Cyanide blocks uptake of oxygen by tissue enzyme systems.
 (e) True Oxygen extraction is increased in stagnant hypoxia.

147 (a) False This causes bronchoconstriction.
 (b) True This causes relaxation of bronchial muscle.
 (c) False Histamine is a bronchoconstrictor.
 (d) True The chromoglycate drugs are thought to work like this and reduce release of bronchoconstrictor agents from mast cells.
 (e) True These suppress bronchoconstrictor and inflammatory mechanisms.

148 (a) True Air flows in since the pressure in the cavity is normally subatmospheric.
 (b) False In ordinary pneumothorax, the intra- and extrapulmonary pressures are equal (atmospheric pressure); collapse is due to elastic recoil of the lungs.
 (c) False It decreases as the lung collapses.
 (d) True The elastic recoil of the lung is no longer applied to the chest wall.
 (e) True The lungs cannot be fully inflated.

149 (a) False There is increased tolerance of high P_{CO_2} levels.
 (b) False Sensitivity to low P_{O_2} remains and is important in maintaining ventilation.
 (c) False HCO_3^- levels rise to compensate for the raised P_{CO_2} in respiratory acidosis.
 (d) False This could arrest ventilation by removing hypoxic drive; 24–28% O_2 would do.
 (e) True When breathing air, $P_{O_2} + P_{CO_2}$ equal about 140 mmHg (19 kPa).

150 (a) False It increases as the walls between alveoli break down to form large sacs.
 (b) True Destruction of elastic fibres holding airways open allows them to narrow.
 (c) False It is increased as airways close more readily than usual.
 (d) True It decreases as the residual volume increases.
 (e) True Thus it is a typical obstructive airways disease.

151 (a) True This 'atelectasis' is due to absorption of trapped air.
 (b) False Local collapse of the lung lowers local intrapleural pressure.
 (c) False Since the affected lung is collapsed it does not count as dead space.
 (d) False Local hypoxia in the unventilated segment causes local vasoconstriction.
 (e) False Vasoconstriction in the collapsed segment prevents deoxygenated blood passing through to the systemic circulation.

152 A shift of the oxygen dissociation curve of blood to the left:
   (a) Decreases the O_2 content of blood at a given P_{O_2}
   (b) Impairs O_2 delivery to the tissues at the normal tissue P_{O_2}
   (c) Occurs in blood perfusing cold extremities
   (d) Occurs in blood stored for several weeks
   (e) Is characteristic of fetal blood when compared with adult blood

153 Obstructive airways disease (OAD) is similar to restrictive lung disease (RLD) in that it reduces:
   (a) Vital capacity (VC)
   (b) The forced expiratory volume in 1 s (FEV_1)
   (c) The ratio FEV_1/VC
   (d) Residual volume
   (e) Peak expiratory flow rate to the same degree

154 A diver breathing air at a depth of 30 m under water:
   (a) Is exposed to a pressure of about four times that at the surface
   (b) Has a raised pressure of nitrogen in the alveoli
   (c) Has a four-fold increase in the oxygen content of blood
   (d) Has a four-fold increase in alveolar water vapour pressure
   (e) Expends less energy than normal on the work of breathing

155 Cyanosis:
   (a) May be caused by high levels of carboxyhaemoglobin in the blood
   (b) May be caused by high levels of methaemoglobin in the blood
   (c) Is seen in fingers of hands immersed in iced water
   (d) Occurs more easily in anaemic than in polycythaemic patients
   (e) Is severe in cyanide poisoning

156 A patient with carbon dioxide retention is likely to have:
   (a) Metabolic acidosis
   (b) Alkaline urine
   (c) Cool extremities
   (d) Raised cerebral blood flow
   (e) Raised plasma bicarbonate

157 Surgical removal of one lung reduces the:
   (a) FEV_1 by about 10%
   (b) Percentage saturation of arterial blood with oxygen
   (c) Exercise tolerance
   (d) Residual volume
   (e) Ventilation/perfusion ratio by about 50%

152 (a) **False** It increases content, especially at tissue P_{O_2} levels.
   (b) **True** The blood does not release its oxygen adequately.
   (c) **True** Cold shifts the curve to the left and may reduce O_2 delivery to cold tissues.
   (d) **True** Such blood when transfused may not release its O_2 content adequately.
   (e) **True** Fetal blood can thus take up O_2 at the low P_{O_2} levels seen in the placenta but fetal tissue P_{O_2} has to be low to permit its release.

153 (a) **True** In OAD by air trapping and in RLD by reducing total lung volume.
   (b) **True** In OAD by slowing flow and in RLD by restricting volume.
   (c) **False** Only OAD which slows flow reduces it.
   (d) **False** It rises in OAD and falls in RLD.
   (e) **False** OAD causes a more severe reduction.

154 (a) **True** 10 m of water = approximately one atmosphere.
   (b) **True** Total alveolar pressure = four atmospheres.
   (c) **False** Haemoglobin is already saturated with O_2 at sea level and cannot take up more; the amount of dissolved oxygen increases, raising total O_2 content by less than 10%.
   (d) **False** This depends on temperature alone.
   (e) **False** More energy is needed to move air made more viscous by compression.

155 (a) **False** Carboxyhaemoglobin is pink and gives the skin a pinkish colour.
   (b) **True** This blue pigment is a rare cause of central cyanosis.
   (c) **False** The fingers are red; cold inhibits oxygen dissociation and reduces metabolism.
   (d) **False** Cyanosis occurs when blood in the skin contains more than 5 g/dl reduced haemoglobin; low haemoglobin values in anaemia make it difficult to reach this level.
   (e) **False** Cyanide poisons the enzymes involved in O_2 uptake by the tissues; in cyanide poisoning the blood remains fully oxygenated and the skin is pink.

156 (a) **False** CO_2 retention causes respiratory acidosis.
   (b) **False** In respiratory acidosis there is increased secretion of H^+ ions in urine.
   (c) **False** Carbon dioxide dilates peripheral blood vessels.
   (d) **True** The vasodilator effect of high P_{CO_2} on cerebral vessels may lead to cerebral oedema and headaches.
   (e) **True** The kidney manufactures bicarbonate to compensate the respiratory acidosis.

157 (a) **False** It reduces it by at least half.
   (b) **False** A single lung can maintain normal oxygenation at rest.
   (c) **True** Maximum ventilation and maximum oxygen uptake are reduced.
   (d) **True** It leads to a restrictive lung disease pattern.
   (e) **False** The ratio is little affected.

158 Coughing:
  (a) Is reflexly initiated by irritation of the alveoli
  (b) Is associated with relaxation of airways smooth muscle
  (c) Depends on contraction of the diaphragm for expulsion of air
  (d) Differs from sneezing in that the glottis is initially closed
  (e) Is depressed during anaesthesia

159 The severity of an obstructive airways disease is indicated by the degree of change in the:
  (a) Total ventilation/perfusion ratio
  (b) Peak expiratory flow rate
  (c) Respiratory quotient
  (d) Tidal volume
  (e) Work of breathing

160 A 50% fall in the ventilation/perfusion ratio in one lung would:
  (a) Lower systemic arterial oxygen content
  (b) Have effects similar to those of a direct right to left atrial shunt
  (c) Increase the physiological dead space
  (d) Lower systemic arterial carbon dioxide content
  (e) Be compensated (with respect to oxygen uptake) by a high ratio in the other lung

161 In the forced expiratory volume (FEV_1) measurement, an adult patient:
  (a) With normal lungs should expire 95% of vital capacity (VC) in 1 s
  (b) With restrictive disease may expire a greater than predicted percentage of VC in the first second
  (c) Who is female would be expected to expire a greater percentage of VC in 1 s than a male of the same age
  (d) With obstructive disease may take more than 5 s to complete the expiration
  (e) With normal lungs should achieve a peak flow rate of at least 200 litres/min

162 The hypoxia in chronic respiratory failure:
  (a) May cause central cyanosis
  (b) May cause peripheral cyanosis
  (c) Leads to increased formation of erythropoietin
  (d) Raises pulmonary vascular resistance
  (e) May lead to right heart failure

163 'Blue bloaters' (patients with chronic obstructive airways disease showing marked cyanosis and oedema) differ from 'pink puffers' (patients with obstructive airways disease showing dyspnoea but not cyanosis) by having a lower:
  (a) Forced expiratory volume in 1 s
  (b) Peak expiratory flow rate
  (c) Arterial blood pH
  (d) Sensitivity to carbon dioxide
  (e) Pulmonary arterial pressure

158 (a) **False** It is initiated by irritation of the trachea and bronchi.
   (b) **False** Contraction narrows airways and increases velocity of flow.
   (c) **False** It depends on expiratory muscles, particularly abdominal muscles.
   (d) **True** Thus it is more explosive than sneezing.
   (e) **True** This may lead to retention of mucous secretions.

159 (a) **False** It may well be normal.
   (b) **True** It is a sensitive indicator.
   (c) **False** RQ is related to the mixture of substrates being metabolized.
   (d) **False** This also may be normal.
   (e) **True** This is related to the increased airway resistance.

160 (a) **True** It constitutes a physiological shunt.
   (b) **True** This would be an anatomical shunt.
   (c) **False** Physiological dead space is relatively overventilated.
   (d) **False** It would tend to raise it.
   (e) **False** Increased ventilation cannot increase oxygen saturation; compensation for carbon dioxide retention can occur.

161 (a) **False** The normal is about 85% at age 20, falling to about 70% at age 60–70.
   (b) **True** The airways are not obstructed and vital capacity volume is reduced.
   (c) **True** Females have, on average, smaller vital capacities than males.
   (d) **True** This is typical of moderately severe obstructive airways disease.
   (e) **True** 200 litres/min is a low figure; most subjects, other than small elderly females, should do better.

162 (a) **True** Reduced V/P ratios allow deoxygenated blood to be shunted to the left side of the heart.
   (b) **False** The peripheral circulation in chronic respiratory failure is usually adequate.
   (c) **True** This in turn leads to the secondary polycythaemia typical of the condition.
   (d) **True** Hypoxia constricts pulmonary vessels and may cause pulmonary hypertension.
   (e) **True** Persistent pulmonary hypertension can lead to right heart failure.

163 (a) **False** This is not a differentiating feature.
   (b) **False** This tends to parallel the FEV_1.
   (c) **True** They have a respiratory acidosis due to CO_2 retention.
   (d) **True** They do not respond adequately to their high P_{CO_2}.
   (e) **False** Pulmonary hypertension in 'blue bloaters' leads to heart failure and oedema.

164 The total amount of O_2 carried by the circulation to the tissues/min (total available oxygen flux):

(a) Normally equals the rate of O_2 consumption by the body/min

(b) Is normally more than 95% combined with haemoglobin

(c) Must fall by about half if haemoglobin concentration is halved

(d) Is more closely related to P_{O_2} than to percentage saturation of the blood with O_2

(e) Must double if body oxygen consumption doubles

164 (a) False At rest, O_2 consumption (about 250 ml/min) is about 25% of the total available.

(b) True All but 3 of the 200 ml/litre is combined with haemoglobin, the rest is dissolved.

(c) False In anaemic hypoxia, cardiac output rises to compensate for the reduced oxygen content per litre of blood.

(d) False If P_{O_2} falls by 25% from normal, there is relatively little change in the blood oxygen content.

(e) False The extraction ratio rises as oxygen consumption rises, e.g. during exercise.

4 ALIMENTARY SYSTEM

165 Bile:
   (a) Contains enzymes required for the digestion of fat
   (b) Contains unconjugated bilirubin
   (c) Salts make cholesterol more water soluble
   (d) Pigments contain iron
   (e) Becomes more alkaline during storage in the gall bladder

166 Saliva:
   (a) From different salivary glands has a similar composition
   (b) Contains enzymes essential for the digestion of carbohydrates
   (c) Has less than half the ionic calcium level of plasma
   (d) Has more than twice the iodide level of plasma
   (e) Has a pH between 5 and 6

167 Swallowing is a reflex which:
   (a) Has its reflex centres in the cervical segments of the spinal cord
   (b) Includes inhibition of respiration
   (c) Is initiated by a voluntary act
   (d) Is dependent on intrinsic nerve networks in the oesophagus
   (e) Is more effective with the trunk in the upright posture

168 Appetite for food is lost when:
   (a) Certain hypothalamic areas are stimulated
   (b) Certain hypothalamic areas are destroyed
   (c) The stomach is distended
   (d) The stomach is surgically removed
   (e) Blood glucose falls

169 Secretion of saliva increases when:
   (a) Touch receptors in the mouth are stimulated
   (b) The mouth is flushed with acid fluids with a pH of about 4
   (c) A subject thinks of unappetizing food
   (d) Vomiting is imminent
   (e) The sympathetic nerve supply is stimulated

165 **(a) False** Bile contains no digestive enzymes; its bile salts assist in the emulsification and absorption of fat.
(b) False The bilirubin is conjugated by the hepatocytes before excretion.
(c) True By forming cholesterol micelles.
(d) False Iron is removed from haem in the formation of bilirubin.
(e) False It becomes more acid, which improves the solubility of bile solids.

166 **(a) False** Serous glands such as the parotids produce a watery juice; mucous glands such as the sublinguals produce a thick viscid juice.
(b) False The functions of salivary amylase (ptyalin) can be effected by enzymes from other digestive glands.
(c) False It is saturated with calcium ions; calcium salts are laid down as plaque.
(d) True Saliva is an important route of iodide excretion; its concentration in saliva is 20–100 times that in plasma.
(e) False Saliva has a neutral pH; acidity in the mouth tends to dissolve tooth enamel.

167 **(a) False** The reflex centres lie in the medulla oblongata.
(b) True This plus closure of the glottis prevents food being aspirated into the airways.
(c) True The voluntary act is the propulsion of a bolus of food onto the posterior pharyngeal wall.
(d) True These are essential for the peristaltic phase.
(e) True Gravity can assist the reflex; tablets stick more when the patient is lying down.

168 **(a) True** For example, when the 'satiety' centres are stimulated.
(b) True For example, when the 'hunger' centres are damaged.
(c) True Appetite is relieved after a meal before the food is absorbed into the blood.
(d) False The drive to eat does not depend on an intact stomach.
(e) False It increases as hypothalamic 'glucostats' detect the low glucose and excite the 'hunger' centres to generate the emotional drive to eat.

169 **(a) True** Food, foreign bodies and the dentist are effective stimuli.
(b) True For example, lemon juice; saliva is a useful buffer to protect teeth from acid.
(c) False Thinking of appetizing food results in salivary secretion by a conditioned reflex; conversely, thinking of unappetizing food inhibits secretion.
(d) True Saliva helps to buffer the acid vomitus when it reaches the mouth.
(e) True Sympathetic stimulation produces a scanty viscid juice; parasympathetic stimulation produces a copious watery juice.

170 Defaecation is a reflex action:
   (a) Which is coordinated by reflex centres in the sacral cord
   (b) Whose afferent limb carries impulses from stretch receptors in the colon
   (c) Whose efferent limb travels mainly in sympathetic autonomic nerves
   (d) Which is more likely to be initiated just after a meal than just before it
   (e) Which can be voluntarily inhibited and facilitated

171 Thirst sensation occurs when:
   (a) Certain areas in the sensory cortex are stimulated
   (b) Blood osmolality is raised but blood volume is normal
   (c) Blood volume is reduced but blood osmolality is normal
   (d) The mouth is dry
   (e) A patient has severe diabetic ketoacidosis

172 In the stomach:
   (a) pH rarely falls below 4.0
   (b) Pepsinogen is converted to pepsin by hydrochloric acid
   (c) Ferrous iron is reduced to ferric iron by hydrochloric acid
   (d) Acid secretion is inhibited by pentagastrin
   (e) There is a rise in the bacterial count after histamine H_1 receptor blockade

173 Intestinal secretions contain:
   (a) Potassium in a concentration similar to that in extracellular fluid
   (b) Enzymes which are released when the vagus nerve is stimulated
   (c) Enzymes which hydrolyse disaccharides
   (d) Enzymes which hydrolyse monosaccharides
   (e) Enzymes which activate pancreatic proteolyic enzymes

174 Pancreatic secretion:
   (a) In response to vagal stimulation is copious, rich in bicarbonate but poor in enzymes
   (b) In response to acid in the duodenum is scanty but rich in enzymes
   (c) In response to secretin is low in bicarbonate
   (d) Contains enzymes which digest neutral fat to glycerol and fatty acids
   (e) Contains enzymes which convert disaccharides to monosaccharides

170 (a) True Reflex defaecation can occur after complete spinal cord transection.
(b) False The stretch receptors triggering the reflex are in the rectal wall.
(c) False Parasympathetic nerves are the main motor nerves for defaecation.
(d) True The wish to defaecate that follows a meal is attributed to a 'gastrocolic reflex'.
(e) True Forced expiration against a closed glottis facilitates the reflex by raising intra-abdominal pressure and pushing faeces into the rectum; contraction of the voluntary muscle of the external sphincter can inhibit it.

171 (a) False The emotional drive to drink depends on 'thirst centres' in the hypothalamus.
(b) True Presumably due to stimulation of hypothalamic osmoreceptors.
(c) True Presumably due to stimulation of vascular volume receptors.
(d) True Afferents from the upper alimentary tract can influence 'thirst centre' activity.
(e) True Probably due to osmoreceptor stimulation by the hyperglycaemic blood and a reduction in blood volume.

172 (a) False Values around pH 2–3 are normal.
(b) True Pepsin is the active proteolyic form of pepsinogen.
(c) False HCl reduces trivalent ferric iron to the divalent ferrous form in which it can be absorbed in the small intestine.
(d) False Pentagastrin is a powerful pharmacological stimulant of mucosal cells to produce HCl.
(e) False But it rises markedly after H_2 blockade which blocks gastric acid secretion.

173 (a) False The concentration is higher, due partly to potassium released from cast off cells.
(b) False Intestinal secretion is not under vagal control; its enzymes are thought to be constituents of the mucosal cells and released when these are cast off into the lumen.
(c) True The mucosal cell brush border contains these enzymes, e.g. maltase and lactase.
(d) False Monosaccharides are end-products of digestion and absorbed as such.
(e) True For example, enterokinase which converts trypsinogen to trypsin.

174 (a) False It is scanty and rich in enzymes; it is secreted reflexly in the 'cephalic' phase of pancreatic secretion when food is thought about or chewed.
(b) False It is copious, bicarbonate rich and poor in enzymes; it buffers the acid secretions entering the duodenum from the stomach.
(c) False This is the hormone released when acid enters the duodenum that stimulates the pancreas to produce the juice described in (b) above.
(d) True Pancreatic lipase.
(e) False Pancreatic amylase breaks down carbohydrates to dextrins and polysaccharides.

175 The liver is the principal site for:
  (a) Synthesis of plasma albumin
  (b) Synthesis of plasma globulins
  (c) Synthesis of vitamin B_{12}
  (d) Storage of vitamin C
  (e) Storage of iron

176 In the colon:
  (a) A greater volume of water is absorbed than in the small intestine
  (b) Mucus is secreted to lubricate the faecal contents
  (c) Faecal transit time is normally about 7 days
  (d) Faecal transit time is inversely related to its fibre content
  (e) Bacteria normally account for about three-quarters of the faecal weight

177 Gastric juice:
  (a) Is secreted when the vagus nerves are stimulated
  (b) Is secreted after the vagus nerves have been cut if food is chewed but not swallowed
  (c) Inactivates the digestive enzymes secreted with saliva
  (d) Is prevented from digesting gastric mucosa by a pepsin inactivator
  (e) Irritates the oesophageal mucosa if regurgitated

178 An increase in body fat increases the:
  (a) Percentage of water in the body
  (b) Survival time during fasting
  (c) Survival time in cold water
  (d) Specific gravity of the body
  (e) Probability of increased morbidity and premature mortality

179 The respiratory quotient:
  (a) Is the ratio of the volume of O_2 consumed to the volume of CO_2 produced
  (b) Depends essentially on the type of substrate being metabolized
  (c) Is 1.0 when glucose is the substrate metabolized
  (d) Is between 0.9 and 1.0 in the second week of fasting
  (e) For the brain is around 1.0

180 Oxygen consumption increases when there is an increase in the:
  (a) Level of oxygen in inspired air
  (b) Metabolic rate
  (c) Body temperature
  (d) Environmental temperature towards thermoneutrality
  (e) Intake of food

175 (a) **True** Plasma albumin concentration falls in liver failure.
(b) **False** Immunoglobulins are made by ribosomes in lymphocytes.
(c) **False** Vitamin B_{12} is ingested in food, absorbed complexed with intrinsic factor in the terminal ileum and stored in the liver.
(d) **False** Vitamin C is not stored; any in excess of requirements is excreted in urine.
(e) **True** Iron released from haem from broken-down RBCs is stored in the liver for future haemopoiesis.

176 (a) **False** The colon absorbs 1–2 litres/day; 8–10 litres are absorbed per day in the small intestine.
(b) **True** Mucous cells are the predominant cells on the colonic mucosal surface.
(c) **False** The average transit time from caecum to pelvic colon is about 12 h but passage from the pelvic colon to the anus may take days.
(d) **True** Fibre (cellulose, lignin, etc.) in the colonic contents stimulates peristaltic movements by adding 'bulk' to the food residues.
(e) **False** But bacteria such as *Escherichia coli* make up about one-third of faecal weight.

177 (a) **True** Vagal stimulation increases acid and pepsinogen secretion; it stimulates mucosal cells by release of acetylcholine and gastrin-releasing peptide at its nerve endings.
(b) **False** After vagotomy, food must enter the stomach to stimulate gastric secretions.
(c) **True** Salivary enzymes are ineffective at the low pH of gastric juice.
(d) **False** But mucosal cells are protected by a coat of mucus impregnated with bicarbonate.
(e) **True** This is normally prevented by the cardiac sphincter.

178 (a) **False** The reverse is true since fat tissue contains little water.
(b) **True** Fat is the main energy store of the body.
(c) **True** It favours survival by increasing skin insulation.
(d) **False** Fat has a lower specific gravity than the lean body mass.
(e) **True** Actuarial tables show this to be true.

179 (a) **False** It is the inverse of this ratio.
(b) **True** It differs for carbohydrate, fat and protein.
(c) **True** $C_6H_{12}O_6 + 6O_2 \rightleftharpoons 6CO_2 + 6H_2O$; each molecule of O_2 consumed results in the production of one molecule of CO_2.
(d) **False** It approaches 0.7, the value when fat is being metabolized.
(e) **True** Carbohydrate is the main substrate for brain metabolism.

180 (a) **False** This is not a determinant of oxygen consumption.
(b) **True** Metabolic rate is the prime determinant of oxygen consumption.
(c) **True** This increases the rate of cellular metabolism.
(d) **False** Less thermogenesis is required to maintain body temperature.
(e) **True** Due to the specific dynamic action of the food, particularly protein.

181 Saliva is necessary for:
  (a) Digestion of food
  (b) Swallowing of food
  (c) Normal speech
  (d) Antisepsis in the mouth
  (e) Taste sensation

182 The stomach:
  (a) Is responsible for absorbing about 10% of the ingested food
  (b) Contains mucosal cells containing high concentrations of carbonic anhydrase
  (c) Peristaltic contractions start from the pyloric region
  (d) Motility increases when fat enters the duodenum
  (e) Relaxes when food is ingested so that there is little rise in intra-gastric pressure

183 Brown fat is:
  (a) More abundant in adults than in infants
  (b) Richer in mitochondria than ordinary fat
  (c) More vascular than ordinary fat
  (d) Stimulated to generate more heat when its parasympathetic nerve supply is stimulated
  (e) More important than shivering in neonatal thermoregulation

184 Nitrogen balance:
  (a) Is the relationship between the body's nitrogen intake and nitrogen loss
  (b) Is positive in childhood
  (c) Becomes more positive when dietary protein is increased
  (d) Becomes negative when a patient is immobilized in bed
  (e) Becomes less negative in the final stages of fatal starvation

185 The normally innervated stomach:
  (a) Is stimulated to secrete gastric juice when food is chewed, even if it is not swallowed
  (b) Cannot secrete HCl when its H_1 histamine receptors are blocked
  (c) And the denervated stomach can secrete gastric juice after a meal is ingested
  (d) Empties more quickly than the denervated stomach
  (e) Is stimulated to secrete gastric juice by the hormone secretin

181 (a) **False** Other digestive tract enzymes can take over if salivary enzymes are absent.
   (b) **False** But swallowing solids is difficult without saliva's moisturizing and lubricant effects.
   (c) **True** Nervous orators with dry mouths continually sip water.
   (d) **True** In the absence of saliva, the mouth becomes infected and ulceration occurs.
   (e) **True** Substances must go into solution before they can stimulate the taste receptors.

182 (a) **False** Little food is absorbed in the stomach other than alcohol.
   (b) **True** This is needed to generate the H^+ ions required for HCl secretion; carbonic anhydrase inhibitors reduce acid secretion.
   (c) **False** They begin at the other end, the fundus.
   (d) **False** Fat in the duodenum inhibits gastric motility.
   (e) **True** This 'receptive relaxation' allows the stomach to store large volumes of food without discomfort.

183 (a) **False** The reverse is true.
   (b) **True** It has a higher metabolic rate than ordinary fat.
   (c) **True** Its higher metabolic rate merits a higher rate of blood flow.
   (d) **False** Metabolic activity in brown fat is stimulated by sympathetic nerve stimulation.
   (e) **True** Infants do not shiver well.

184 (a) **True** It is positive when more nitrogen is taken in than is lost.
   (b) **True** Nitrogen intake is greater than nitrogen loss during active tissue growth.
   (c) **False** The additional nitrogen intake is balanced by additional nitrogen loss when the extra protein is metabolized and its nitrogen excreted in the urine.
   (d) **True** Muscles waste and the protein released is metabolized.
   (e) **False** It becomes more negative when little but protein remains to be metabolized.

185 (a) **True** This depends on a vagal reflex.
   (b) **False** Blockade of histamine H_2 receptors blocks gastric acid secretion.
   (c) **True** The denervated stomach is still affected by gastrin released from the gastric mucosa when food enters the stomach.
   (d) **True** Gastric motility is more effective when it is coordinated by vagal nerve activity.
   (e) **False** Secretin, which stimulates pancreatic secretion, decreases gastric secretion.

186 The passage of gastric contents to the duodenum may cause:
   (a) Copious secretion of pancreatic juice rich in bicarbonate
   (b) Decreased gastric motility
   (c) Contraction of the gall bladder
   (d) Contraction of the sphincter of Oddi
   (e) Release of pancreozymin

187 Bile salts:
   (a) Are the only constituents of bile necessary for digestion
   (b) Have a characteristic molecule, part water-soluble and part fat-soluble
   (c) Are reabsorbed mainly in the upper small intestine
   (d) Are derived from cholesterol
   (e) Inhibit bile secretion by the liver

188 Absorption of:
   (a) Fat is impaired if either bile or pancreatic juice is not available
   (b) Undigested protein molecules can occur in the newborn
   (c) Laevo-amino acids occurs more rapidly than absorption of the dextro
       forms
   (d) Iron is at a rate proportional to body needs
   (e) Sodium is at a rate proportional to body needs

189 The specific dynamic action of food:
   (a) Is the increase in metabolic rate that results from eating food
   (b) Persists for about an hour after a meal is ingested
   (c) Is due to the additional energy expended in digesting and absorbing the
       food
   (d) Results in about 30% of the energy value of ingested protein being
       unavailable for other purposes
   (e) Results in about 20% of the energy value of ingested fat and carbohydrate
       being unavailable for other purposes

190 Secretion of gastric juice:
   (a) Increases when food stimulates mucosal cells in the pyloric region
   (b) Is associated with a decrease in the pH of venous blood draining the
       stomach
   (c) In response to food is reduced if the vagus nerves are cut
   (d) Is essential for protein digestion
   (e) Is essential for absorption of vitamin B_{12}

191 In the small intestine:
   (a) The enzyme concentration in intestinal juice is lower in the ileum than in
       the jejunum
   (b) Vitamin B_{12} is absorbed mainly in the jejunum
   (c) Water absorption depends on the active absorption of sodium and glucose
   (d) Absorption of calcium occurs mainly in the terminal ileum
   (e) Glucose absorption is independent of sodium absorption

186 (a) **True** This is caused by the hormone secretin released from the duodenal mucosa.
(b) **True** This postpones further gastric emptying.
(c) **True** Due to the action of cholecystokinin released from the mucosal cells.
(d) **False** The sphincter of Oddi must relax to allow bile to enter the gut.
(e) **True** Pancreozymin stimulates the pancreas to secrete a scanty, enzyme-rich juice.

187 (a) **True** They emulsify fat, creating a greater surface area for lipase to act on.
(b) **True** This property allows them to form micelles for fat transport.
(c) **False** They are absorbed in the terminal ileum.
(d) **True** They are synthesized from cholesterol in the liver.
(e) **False** They are 'choleretics', substances which stimulate bile secretion.

188 (a) **False** If either is absent, undigested fat appears in the faeces.
(b) **True** Maternal antibodies (globulins) in colostrum are so absorbed.
(c) **True** Transport mechanisms are isomer-specific.
(d) **True** An active carrier-mediated transport mechanism is involved which reduces the danger of iron toxicity with excessive intake.
(e) **False** Most ingested sodium is absorbed; the total body sodium content is controlled by hormonal and renal mechanisms.

189 (a) **True** This is its definition.
(b) **False** It usually lasts about 6 h.
(c) **False** It includes energy expended on processing absorbed material for detoxication, metabolism and storage.
(d) **True** Mainly because of the energy required to deaminate amino acids.
(e) **False** The figures are around 4% for fat and 6% for carbohydrate.

190 (a) **True** Cells in the pyloris produce gastrin when food enters the stomach.
(b) **False** Venous blood pH rises as bicarbonate enters the circulation in the 'alkaline tide'.
(c) **True** Vagal activity plays an important role in such secretion.
(d) **False** Pancreatic trypsin and chymotrypsin can digest proteins.
(e) **True** Without gastric intrinsic factor, Vitamin B_{12} is not absorbed from the gut.

191 (a) **True** Being proteins, they are digested by proteolytic enzymes in the gut
(b) **False** It is absorbed mainly in the terminal ileum.
(c) **True** Water is absorbed passively down the osmotic gradient set up by active sodium and glucose absorption.
(d) **False** It occurs mainly in the duodenum.
(e) **False** Sodium is required at the luminal surface for glucose to be absorbed by an active carrier-mediated process.

192 The cells of the liver:
    (a) Help to maintain the normal blood glucose level
    (b) Deaminate amino acids to form NH_4^+ which is excreted as ammonium salts in the urine
    (c) Synthesize vitamin D_3 (cholecalciferol)
    (d) Manufacture immune globulins
    (e) Inactivate gonadal steroid hormones

193 Absorption of dietary fat:
    (a) Occurs only after the neutral fat has been split into glycerol and fatty acids
    (b) Involves fat uptake by both the lymphatic and blood capillaries
    (c) Is impaired following gastrectomy
    (d) Is required for normal bone development
    (e) Is required for normal blood clotting

194 One gram of:
    (a) Carbohydrate, metabolized in the body, yields the same energy as when oxidized in a bomb calorimeter
    (b) Fat, metabolized in the body, yields 10% more energy than 1 g of carbohydrate
    (c) Protein, metabolized in the body, yields the same energy as when oxidized in a bomb calorimeter
    (d) Carbohydrate, metabolized in the body, yields about the same energy as 1 g of protein
    (e) Protein per kg body weight is an adequate daily protein intake for a sedentary adult

195 Cholesterol:
    (a) Can be absorbed from the gut by incorporation into chylomicrons taken up by lymphatics
    (b) Can be synthesized in the liver
    (c) In the diet comes from animal and vegetable sources
    (d) Is eliminated from the body mainly by metabolic degradation
    (e) Is a precursor of adrenal cortical hormones

196 Free (non-esterified) fatty acids in plasma:
    (a) Account for less than 10% of the total fatty acids in plasma
    (b) Are complexed with the plasma proteins
    (c) Decrease when the level of blood adrenaline rises
    (d) Can be metabolized to release energy in cardiac and skeletal muscle
    (e) Can be metabolized to release energy in the brain

192 (a) True When blood glucose falls, liver glycogen is broken down to form glucose; when glucose levels rise above normal, glucose is taken up by the liver and stored as liver glycogen.

(b) False The NH_4^+ is converted into urea and excreted in the urine; NH_4^+ is toxic.

(c) False Cholecalciferol in produced in skin by the action of sunlight; the liver converts it to 25-hydroxycholecalciferol and the kidney completes its activation by further hydroxylation.

(d) False They manufacture most of the plasma proteins but lymphocytes manufacture immune globulins.

(e) True The failure to inactivate oestrogens in men with liver failure can lead to breast enlargement.

193 (a) False Unsplit neutral fat can be absorbed if emulsified into sufficiently small particles.

(b) True The smaller fatty acids pass directly into blood; the larger ones are esterified, packaged into chylomicrons and taken into lymphatics.

(c) True The loss of gastric storage capacity results in rapid transit of food through the small intestine and insufficient time for complete digestion and absorption of fat.

(d) True Vitamin D is required for normal bone development; it is a fat-soluble vitamin and absorbed along with the fat.

(e) True Vitamin K, which is needed for the synthesis of certain clotting factors in the liver, is also a fat-soluble vitamin.

194 (a) True In both cases the carbohydrate is oxidized to carbon dioxide and water.

(b) False Fat yields approximately 125% more energy.

(c) False It yields about 40% more in the calorimeter; in the body, urea derived from protein is excreted in the urine and its free energy is not released to the body.

(d) True The free energy released by carbohydrate and protein in the body is similar.

(e) True This is probably a high figure; protein requirements increase with increasing levels of energy expenditure.

195 (a) True Most of it is incorporated into very low density lipoproteins (VLDL) and circulates as such.

(b) True Its synthesis is determined mainly by saturated fat intake.

(c) False The main sources are egg yolk and animal fat; there is no cholesterol in vegetables.

(d) False It is eliminated mainly by excretion in the bile.

(e) True It is a precursor of a number of steroid hormones.

196 (a) True Most are esterified to glycerol and cholesterol.

(b) True This is necessary for solubility.

(c) False They rise due to release from adipose tissue.

(d) True They are an important source of energy for contraction.

(e) False Glucose is the only normal substrate for energy production here.

197 The risk of developing gallstones increases:
- (a) When cholesterol micelles are formed in the gall bladder
- (b) As the bile salt:cholesterol ratio increases
- (c) As the lecithin:cholesterol ratio increases
- (d) When supplementary bile salts are taken by mouth
- (e) In patients with haemolytic anaemia

198 Modifying gastric function:
- (a) By cutting vagal fibres to the pylorus increases the emptying rate
- (b) By enlarging the pyloric orifice increases the emptying rate
- (c) By making a side-to-side anastomosis between the stomach and jejunum may cause peptic ulceration in the jejeunum
- (d) To allow rapid gastric emptying may lead to low blood glucose levels after a meal
- (e) To allow rapid gastric emptying may lead to a fall in blood volume and blood pressure after a large meal

199 Impaired intestinal absorption of:
- (a) Iron occurs frequently following removal of most of the stomach
- (b) Iodide leads to a reduction in size of the thyroid gland
- (c) Water occurs in infants who cannot digest disaccharides
- (d) Calcium may occur following removal of the terminal ileum
- (e) Bile salts may occur following removal of the terminal ileum

200 Peptic ulcer pain is typically relieved by:
- (a) Raising the pH of the fluid bathing the ulcer
- (b) Oral administration of sodium bicarbonate
- (c) Oral administration of a non-absorbable antacid
- (d) A drug which interferes with the action of acetylcholinesterase
- (e) A drug which blocks the gastric proton pump

201 Fat stores in the adult:
- (a) Make up less than 5% of average body weight
- (b) Make up a smaller percentage of body weight in women than in men
- (c) Release fatty acids when there is increased sympathetic nerve activity
- (d) Release fatty acids when insulin is injected
- (e) Enlarge by increasing the number of adipocytes they contain

202 Metabolic rate can be estimated from measurements of:
- (a) Total heat production
- (b) The calorific value of the food consumed in the previous 24 h
- (c) Oxygen consumption, provided the type of food being metabolized is known
- (d) Oxygen consumption and the respiratory quotient
- (e) Carbon dioxide production and the respiratory quotient

197 (a) **False** This is a vital solubilizing mechanism.
   (b) **False** This favours formation of micelles.
   (c) **False** Lecithin also contributes to the formation of micelles.
   (d) **False** This can decrease the risk temporarily.
   (e) **True** Excess bilirubin can give rise to pigment stones.

198 (a) **False** Emptying is slowed as vagal coordination of gastric motility is lost.
   (b) **True** This may compensate for the slowing effects of vagotomy.
   (c) **True** Acid-pepsin can cause a jejunal stomal ulcer.
   (d) **True** Rapid glucose absorption leads to excessive insulin secretion and a consequent fall in blood glucose.
   (e) **True** In this further variety of the 'dumping syndrome', the sudden arrival of many osmotically active particles in the intestine draws extracellular fluid into the gut.

199 (a) **True** Due to loss of gastric acid which reduces ferric to the ferrous iron, the form in which it is absorbed. The other problem affecting iron absorption is rapid intestinal transit.
   (b) **False** The size increases due to increased TSH stimulation.
   (c) **True** The unabsorbed disaccharides cause osmotic diarrhoea.
   (d) **False** Calcium is absorbed mainly in the duodenum.
   (e) **True** The terminal ileum is the main site of bile salt absorption.

200 (a) **True** Acid is the cause of pain in these ulcers.
   (b) **True** But this is absorbed and may cause alkalosis.
   (c) **True** This also reduces acidity, but for a short time only.
   (d) **False** This would increase vagal acid production.
   (e) **True** This can give long-lasting relief.

201 (a) **False** The normal value is around 15–20%.
   (b) **False** Fat, as a percentage of body weight, is 5–10% higher in women than in men.
   (c) **True** This is mediated via beta (β) adrenergic receptor stimulation.
   (d) **False** Insulin favours deposition of fat in the fat stores.
   (e) **False** In obesity, adipocytes increase in size rather than number.

202 (a) **True** Total energy expenditure must eventually appear as heat.
   (b) **False** Energy production is related to the metabolism of food, not its intake.
   (c) **True** Since oxygen is consumed only in metabolism, total oxygen consumption is an index of metabolic rate; the energy produced per unit of oxygen consumed varies somewhat with different substrates.
   (d) **True** Respiratory quotient indicates the mix of substrates used.
   (e) **True** Metabolic rate is proportional to carbon dioxide formation.

203 Intestinal obstruction causes:
   (a) Constipation
   (b) Crampy pain due to intermittent vigorous peristalsis
   (c) Distension due to fluid and gas proximal to the obstruction
   (d) Hypotension
   (e) Vomiting which is more severe with low than with high bowel obstruction

204 The cause of jaundice is likely to be:
   (a) Liver disease, if albumin is low and bilirubin mainly unconjugated
   (b) Bile duct obstruction, if the urine is paler than normal
   (c) Haemolysis, if the prothrombin level is below normal
   (d) Haemolysis, if the urine is darker than normal
   (e) A post-hepatic cause, if the bilirubin level is high

205 Complications that may arise after total gastrectomy include:
   (a) Inadequate food intake
   (b) Depletion of vitamin B_{12} stores in the liver
   (c) Malabsorption of fat due to rapid intestinal transit
   (d) Impaired defaecation due to loss of the gastrocolic reflex
   (e) Inability to digest protein

206 Severe diarrhoea causes a decrease in:
   (a) Body potassium
   (b) Body sodium
   (c) Extracellular fluid volume
   (d) Total peripheral resistance
   (e) Blood pH

207 Diminished liver function may result in an increase in the:
   (a) Albumin/globulin ratio in the blood
   (b) Breast size in the male
   (c) Level of unconjugated bilirubin in the blood
   (d) Tendency to bleed
   (e) Variability of the blood glucose level

208 Peptic ulceration tends to heal after:
   (a) Division of the vagal nerve supply to the stomach
   (b) Surgical removal of the pyloric antrum
   (c) Glucocorticoid drug treatment
   (d) Aspirin therapy
   (e) Treatment with histamine H_2 blocking drugs

203 (a) **True**   This is complete when the bowel below the block is empty.
   (b) **True**   Intermittent colic is typical of early obstruction.
   (c) **True**   This may be severe when the obstruction is low down.
   (d) **True**   Due to hypovolaemia as secretions accumulate above the obstruction.
   (e) **False** Vomiting is worse with high obstruction since the copious upper alimentary secretions cannot be absorbed.

204 (a) **True**   Both indicate impairment of liver function.
   (b) **False** Urine is dark in jaundice due to bile duct obstructions since conjugated bilirubin can pass the glomerular filter.
   (c) **False** A low prothrombin suggests impaired liver function.
   (d) **False** Haemolysis raises the level of unconjugated bilirubin which is protein bound and not excreted in the urine (acholuric jaundice).
   (e) **False** It is high with all causes of jaundice.

205 (a) **True**   Due to loss of the 'reservoir' function of the stomach.
   (b) **True**   Due to loss of gastric 'intrinsic factor', required for absorption of the vitamin.
   (c) **True**   The mechanism regulating food delivery to the small intestine is lost.
   (d) **False** This 'reflex' is not essential for defaecation.
   (e) **False** Pancreatic enzymes can make up for the loss of pepsin.

206 (a) **True**   Due to loss of secretions rich in potassium.
   (b) **True**   Sodium is the main cation lost in diarrhoea.
   (c) **True**   Body sodium is the 'skeleton' of extracellular fluid volume.
   (d) **False** This is raised to maintain arterial blood pressure.
   (e) **True**   Loss of intestinal bicarbonate causes metabolic acidosis.

207 (a) **False** Albumin, which is manufactured in the liver, falls; globulin tends to rise.
   (b) **True**   Due to loss of the liver's ability to conjugate oestrogens and progestogens.
   (c) **True**   The liver normally conjugates bilirubin with glucuronic acid.
   (d) **True**   Due to reduced synthesis of clotting factors by the liver.
   (e) **True**   Due to the liver's reduced ability to store and mobilize glycogen.

208 (a) **True**   This reduces gastric acidity.
   (b) **True**   Gastrin, which stimulates acid secretion, is produced in the pyloric antrum.
   (c) **False** Glucocorticoids may make the ulceration worse by inhibiting tissue repair.
   (d) **False** This aggravates ulceration by impairing secretion of protective mucus.
   (e) **True**   These reduce acidity and are widely used in the treatment of peptic ulceration.

209 Urobilinogen is:
(a) A mixture of colourless compounds also known as stercobilinogen
(b) Formed in the reticuloendothelial system from bilirubin
(c) Converted into the dark pigment, urobilin, on exposure to air
(d) Absorbed from the intestine
(e) Excreted mainly in the urine

210 Surgical removal of 90% of the small intestine may cause a decrease in:
(a) The fat content of the stools
(b) Bone mineralization (osteomalacia)
(c) Extracellular fluid volume
(d) Blood haemoglobin level
(e) Body weight

211 Lack of pancreatic juice in the duodenum may lead to:
(a) The presence of undigested meat fibres in the stools
(b) An increase in the fat content of the faeces
(c) Faeces having a high specific gravity
(d) A tendency for the faeces to putrefy
(e) A high prothrombin level in blood

212 In liver failure there is likely to be:
(a) Salt and water retention
(b) Dependent oedema
(c) Raised blood urea
(d) Impaired absorption of fat
(e) Intoxication after eating a high-protein meal

213 Gastric:
(a) Acid secretion in response to low blood sugar is mediated by the hormone gastrin
(b) Emptying is facilitated by activity in sympathetic nerves
(c) Acid secretion increases when histamine H_2, muscarinic M_1 or gastrin receptors are activated
(d) Acid secretion is inhibited by the presence of food in the duodenum
(e) Contraction waves pass over the stomach at a rate of about ten per minute

214 In portal hypertension:
(a) The total vascular resistance of the hepatic sinusoids is increased
(b) Portal blood flow through the liver is increased
(c) The volume of fluid in the peritoneal cavity increases
(d) A porto-caval shunt (anastomosis between portal vein and inferior vena cava) can decrease bleeding into the alimentary tract
(e) A porto-caval shunt increases the risk of coma after bleeding into the alimentary tract

209 (a) True It is present in both urine and faeces.
(b) False It is formed in the intestine.
(c) True Urine rich in urobilinogen (due to haemolysis) darkens on standing.
(d) True And carried back to the liver in the enterohepatic circulation.
(e) False It is excreted mainly in the bile.

210 (a) False Fat absorption is incomplete and this causes steatorrhoea.
(b) True There is poor absorption of fat-soluble vitamin D.
(c) False Adequate salt and water absorption still occur.
(d) True Vitamin B_{12} and iron absorption are impaired and this may cause anaemia.
(e) True Due to the malabsorption of food.

211 (a) True Due to lack of the proteinases trypsin and chymotrypsin.
(b) True Due to lack of pancreatic lipase.
(c) False A high fat content lowers specific gravity; stools may float.
(d) True Due to the high level of nutrients in the faeces, bacteria flourish.
(e) False It is reduced; malabsorption of fat reduces absorption of fat-soluble vitamin K.

212 (a) True Due in part to the liver's failure to conjugate salt-retaining hormones such as aldosterone.
(b) True Due to the above plus depletion of plasma albumin.
(c) False It falls due to impaired urea synthesis from NH_4^+ released in deamination.
(d) True Due to impaired formation and excretion of bile salts.
(e) True Due to inadequate detoxification of toxins derived from proteins in the diet.

213 (a) False It is reflexly mediated by the vagus and is absent in vagotomized stomachs.
(b) False This delays gastric emptying.
(c) True Histamine from mast-like cells activates H_2 receptors, acetylcholine from parasympathetic nerve endings activates M_1 receptors and gastrin activates gastrin receptors.
(d) True The neural/hormonal mechanisms responsible for the inhibition are not known.
(e) False The normal rate is around three per minute.

214 (a) True This is the cause of the hypertension.
(b) False It is decreased as blood is diverted to alternative routes back to the great veins.
(c) True Raised hydrostatic pressure increases filtration from visceral capillaries; fluid accumulates in the peritoneal cavity to cause ascites.
(d) True By diverting blood away from oesophageal varices.
(e) True The failure of the liver to detoxicate toxic end-products of protein metabolism in liver failure can lead to hepatic encephalopathy; bypassing the liver with a porto-caval shunt may aggravate the condition, especially after an intestinal bleed.

215 Constipation is a recognized consequence of:
   (a) Sensory denervation of the rectum
   (b) Psychological stress
   (c) Abnormality of the autonomic nerve supply to the colon
   (d) A diet which leaves little unabsorbed residue in the gut
   (e) Overactivity of the thyroid gland

216 Absorption of glucose by intestinal mucosal cells:
   (a) Relies on a carrier mechanism in the cell membrane
   (b) Is blocked by the same agents which block renal reabsorption of glucose
   (c) Is enhanced by blockade of active sodium transport in the cells
   (d) Involves the same carriers that are used for the absorption of galactose
   (e) Normally takes place mainly in the ileum

217 Vomiting:
   (a) Leads to expulsion of gastric contents by violent rhythmic contractions of gut smooth muscle
   (b) Is coordinated by a mid-brain vomiting centre
   (c) Of green fluid suggests that duodenal contents have regurgitated into the stomach
   (d) May be accompanied by a fall in arterial blood pressure
   (e) May be induced by drugs acting on centres in the medulla

218 Obesity:
   (a) Is unlikely to occur on a high-protein diet even if the calorific value of the food exceeds daily energy expenditure
   (b) Is associated with increased demands on pancreatic islet beta cells
   (c) Can be assessed by multiple measurements of skin-fold thickness
   (d) Can be assessed by weighing the body in air and in water
   (e) Is not present until body weight is 40% above normal

219 Muscle tone in the lower oesophagus is:
   (a) Greater than tone in the middle oesophagus
   (b) A major factor in preventing heartburn
   (c) Increased in pregnancy
   (d) Increased by gastrin
   (e) Increased by anticholinergic drugs

215 (a) **True**   This breaks the reflex arc on which defaecation depends.
   (b) **True**   Stress may modify the reflex to cause either constipation or diarrhoea.
   (c) **True**   In children this may cause megacolon.
   (d) **True**   Frequency of defaecation is related to the bulk of food residues.
   (e) **False**   This leads to increased frequency of defaecation.

216 (a) **True**   This facilitates transport from the lumen.
   (b) **True**   Phloridzin has this action; the carrier mechanisms at the two sites are similar.
   (c) **False**   It is impaired, suggesting that sodium absorption facilitates glucose absorption.
   (d) **True**   But different carriers are involved in fructose absorption.
   (e) **False**   It takes place mainly in the duodenum and jejunum.

217 (a) **False**   Forced expiratory efforts by skeletal muscles in the presence of a closed glottis and pylorus are responsible for compressing the stomach.
   (b) **False**   It is coordinated in the medulla oblongata.
   (c) **True**   Bile pigments enter the gut in the duodenum.
   (d) **True**   This is one of several types of associated autonomic disturbance.
   (e) **True**   Apomorphine is an emetic drug which acts on a chemoreceptor trigger zone for vomiting.

218 (a) **False**   If food energy intake exceeds energy expenditure, obesity develops regardless of the predominant food in the diet.
   (b) **True**   Mild maturity-onset diabetes mellitus may be relieved by loss of body weight.
   (c) **True**   Subcutaneous fat is a good indicator of its severity.
   (d) **True**   Body specific gravity which is inversely related to body fat content can be calculated from these values.
   (e) **False**   The conventional threshold is 10%.

219 (a) **True**   The high-pressure zone indicates the 'cardiac sphincter'.
   (b) **True**   It prevents reflux of acid into the oesophagus.
   (c) **False**   It is decreased and heartburn is common in pregnancy.
   (d) **True**   This prevents reflux during gastric contractions.
   (e) **False**   These reduce tone, suggesting that cholinergic nerves have a role in maintaining normal tone.

5 NEUROMUSCULAR SYSTEM

220 A reflex action:
    (a) Is initiated at a sensory receptor organ
    (b) May result in endocrine secretion
    (c) Involves transmission across at least two central nervous synapses in series
    (d) May be excitatory or inhibitory
    (e) Is independent of higher centres in the brain

221 In skeletal muscle neuromuscular junctions:
    (a) The motor end-plate is the motor nerve terminal
    (b) Spontaneous (miniature) potentials may be recorded in the motor nerve terminal
    (c) Motor nerve terminals have vesicles containing acetylcholine
    (d) There is a high concentration of acetylcholinesterase
    (e) Transmission is facilitated by botulinum toxin

222 Cerebrospinal fluid:
    (a) Is formed in the arachnoid granulations
    (b) Provides the brain with most of its nutrition
    (c) Protects the brain from injury when the head is moved
    (d) Has a lower pressure than that in the cerebral venous sinuses
    (e) Flow around the adult brain is around 0.5 litre/day

223 A skeletal muscle fibre:
    (a) Membrane is negatively charged on the inside with respect to the outside at rest
    (b) Contains intracellular stores of calcium ions
    (c) Is normally innervated by more than one motor neurone
    (d) Becomes more excitable as its resting membrane potential falls
    (e) Becomes less excitable as the extracellular ionized calcium levels fall

224 In sensory receptors:
    (a) Stimulus energy is converted into a local depolarization
    (b) The generator potential is graded and self-propagating
    (c) A generator potential can be produced by only one form of energy
    (d) The frequency of action potentials generated doubles when the strength of the stimulus doubles
    (e) Serving touch sensation, constant suprathreshold stimulation causes action potentials to be generated at a constant rate

220 (a) True   Stimulation of the receptor generates impulses in the afferent limb.

   (b) True   Stimulation of osmoreceptors reflexly modifies ADH output from the posterior pituitary gland.

   (c) False   The knee jerk reflex arc contains only one central nervous synapse.

   (d) True   In the knee jerk. extensors of the knee contract, while flexors relax (reciprocal inhibition).

   (e) False   Higher centres can facilitate or inhibit many reflex actions such as micturition.

221 (a) False   It is the modified muscle membrane adjacent to the nerve terminal.

   (b) False   They may be recorded at the motor end-plate.

   (c) True   This neurotransmitter, released by exocytosis, excites the end-plate membrane.

   (d) True   This makes acetylcholine's action transient.

   (e) False   Botulinum toxin blocks transmission by an action on the motor nerve terminals.

222 (a) False   It is formed in choroid plexuses by active and passive processes.

   (b) False   Most of the brain's nutrition comes from the blood.

   (c) True   This is its main function and it does so by providing cushioning and buoyancy.

   (d) False   Its higher pressure allows drainage by filtration to the dural venous sinuses via the arachnoid villi.

   (e) True   This is about four times its volume.

223 (a) True   This 'resting membrane potential' is about -90 mV.

   (b) True   These are released on excitation.

   (c) False   A single neurone supplies a group of muscle fibres.

   (d) True   It becomes more excitable as its membrane potential approaches the firing threshold (about -70 mV).

   (e) False   Decreasing extracellular Ca^{2+} increases excitability and may lead to spontaneous contractions (tetany), possibly by increasing sodium permeability.

224 (a) True   This is the generator potential.

   (b) False   It is graded but not propagated.

   (c) False   But sensitivity is greatest to one form of energy – the 'adequate stimulus'.

   (d) False   Frequency is related to the logarithm of the strength of the stimulus.

   (e) False   Frequency falls with time due to adaptation.

225 A somatic lower motor neurone:
  (a) Innervates fewer fibres in an eye muscle than does a neurone innervating a leg muscle
  (b) Conducts impulses at a speed similar to that in an autonomic postganglionic neurone
  (c) Is unmyelinated
  (d) Conducts impulses which cause relaxation in some skeletal muscles
  (e) Synapses with skeletal muscle but not with other neurones

226 Impulses serving pain sensation in the left foot are relayed:
  (a) Across synapses in the left posterior root ganglion
  (b) By fibres in the left spinothalamic tract
  (c) By the same spinal cord tract which serves heat and cold sensation
  (d) To the thalamus on the right side
  (e) To the cerebral cortex before entering consciousness

227 An excitatory postsynaptic potential:
  (a) Is the depolarization of a postsynaptic nerve cell membrane that occurs when a presynaptic neurone is stimulated
  (b) Involves reversal of polarity across the postsynaptic nerve cell membrane
  (c) May be recorded from a posterior root ganglion cell
  (d) Is propagated at the same rate as an action potential
  (e) Is caused by the electrical field induced by activity in the presynaptic nerve terminals

228 The ascending reticular formation:
  (a) When stimulated tends to increase alertness
  (b) Transmits impulses to higher centres via a multisynaptic pathway
  (c) Is activated by collateral branches of sensory neurones
  (d) Neurones project to most parts of the cerebral cortex
  (e) Increases its activity during deep sleep

229 The cerebellum:
  (a) Modifies the discharge of spinal motor neurones
  (b) Is essential for finely coordinated movements
  (c) Has an afferent input from the motor cortex
  (d) Has an afferent input from muscle proprioceptors
  (e) Has an afferent input from the vestibular system

230 During deep sleep there is a fall in:
  (a) Hand skin temperature
  (b) Arterial P_{CO_2}
  (c) Blood growth hormone/cortisol ratio
  (d) Metabolic rate
  (e) Urine formation

225 (a) **True**   The more precise the movement required, the fewer the fibres supplied by one motor neurone.
   (b) **False**   Somatic motor neurones conduct at 60–120 m/s; autonomic at about 1 m/s.
   (c) **False**   Fast-conducting fibres are large and myelinated.
   (d) **False**   Impulses carried by somatic motor neurones are excitatory to skeletal muscle.
   (e) **False**   Some carry impulses to inhibitory (Renshaw) cells in the anterior horn.

226 (a) **False**   The ganglion contains sensory cell bodies but no synapses; primary pain fibres carrying these impulses synapse with secondary neurones in the left posterior horn.
   (b) **False**   They cross to the right spinothalamic tract.
   (c) **True**   Pain and temperature fibres travel together.
   (d) **True**   Sensations other than smell are relayed in the thalamus.
   (e) **False**   They enter consciousness at a subcortical level.

227 (a) **True**   An inhibitory potential is a hyperpolarization.
   (b) **False**   It is a transient (about 5 ms), small (about 5 mV) depolarization towards the threshold for firing.
   (c) **False**   It may be recorded from a motor neurone.
   (d) **False**   It is not propagated.
   (e) **False**   It is caused by transmitters changing permeability in postsynaptic membranes.

228 (a) **True**   Associated with increased electrical activity in cortical regions.
   (b) **True**   Diffuse cortical activity following sensory stimulation occurs later than local postcentral gyrus activity.
   (c) **True**   Sensory fibres ascending to the thalamus send collaterals to reticular nuclei.
   (d) **True**   The diffuse cortical activity following sensory stimulation is abolished by damage to the reticular formation.
   (e) **False**   Decreased activity leads to a higher arousal threshold during sleep.

229 (a) **True**   Hence it influences skeletal muscle activity.
   (b) **True**   Damage to it leads to movements being clumsy (ataxia).
   (c) **True**   This provides information on the desired movement.
   (d) **True**   This provides information on the actual movement.
   (e) **True**   It has an important role in maintaining balance.

230 (a) **False**   Skin temperature rises due to vasodilation.
   (b) **False**   The level rises due to hypoventilation.
   (c) **False**   Growth hormone level is higher, cortisol lower in sleep.
   (d) **True**   Due to slowing of metabolism and a fall in catecholamine level.
   (e) **True**   The antidiuretic hormone level is higher in sleep.

231 Sympathetic:
  (a) Ganglionic transmission is mediated by acetylcholine
  (b) Neuromuscular transmission at the heart is mediated by adrenaline
  (c) Neuromuscular transmission in skin arterioles is mediated by acetylcholine
  (d) Neuroglandular transmission at sweat glands is mediated by noradrenaline
  (e) Neuromuscular transmission at the iris is mediated by noradrenaline

232 The blood–brain barrier:
  (a) Slows equilibration of solutes between blood and brain tissue fluids
  (b) Is a more effective barrier for fat-soluble substances than water-soluble substances
  (c) Is a more effective barrier in infants than in adults
  (d) Is a more effective barrier for CO_2 than for O_2
  (e) Permits hydrogen ions to pass freely

233 Nerve impulses:
  (a) Can travel in one direction only in a nerve fibre
  (b) Can travel in one direction only across a synapse
  (c) Travel at the speed of an electric current
  (d) Correspond in duration to that of the nerve refractory period
  (e) Can be transmitted at higher frequencies in autonomic than in somatic nerves

234 In skeletal muscle:
  (a) Contraction occurs when its pacemaker cells depolarize sufficiently to reach the threshold for firing
  (b) Calcium is taken up by the sarcotubular system when it contracts
  (c) Actin and myosin filaments shorten when it contracts
  (d) The sarcomeres shorten during contraction
  (e) Contraction strength is related to initial length of the muscle fibres

235 The electroencephalogram normally shows voltage waves:
  (a) Whose amplitude is related to intelligence
  (b) Of smaller amplitude during deep sleep than during alert wakefulness
  (c) Of lower frequency during deep sleep than during alert wakefulness
  (d) Of greater amplitude than those of the electrocardiogram
  (e) Which are bilaterally symmetrical

236 Saltatory conduction:
  (a) Occurs only in myelinated fibres
  (b) Does not depend on depolarization of the nerve membrane
  (c) Has a slower velocity in cold than in warm conditions
  (d) Is faster than non-saltatory conduction in nerve fibres with diameters around 10 μm
  (e) Transmits impulses with a velocity proportional to fibre diameter

231 (a) **True** This applies to parasympathetic ganglia also.
(b) **False** The transmitter is noradrenaline.
(c) **False** The transmitter is noradrenaline here also.
(d) **False** These sympathetic fibres are cholinergic.
(e) **True** Sympathetic adrenergic fibres innervate radial muscle in the iris; cholinergic parasympathetic fibres innervate the circular muscle.

232 (a) **True** Capillary permeability is lower in the brain than in other tissues.
(b) **False** Fat-soluble substances cross readily due to the lipid in cell membranes.
(c) **False** The reverse is true; bilirubin, which cannot pass in the adult, does so in infants.
(d) **False** Both CO_2 and O_2 cross the blood–brain barrier easily.
(e) **False** This protects brain tissues against sudden changes in the pH of blood.

233 (a) **False** In axons, impulses travel in both directions from a point of electrical stimulation.
(b) **True** The transmitter vesicles are in the presynaptic terminal.
(c) **False** Impulse propagation is a different process and much slower than electrical current.
(d) **True** Nerves cannot be re-excited while their membrane polarity is reversed.
(e) **False** The shorter refractory periods in somatic nerves allows higher frequencies.

234 (a) **False** Skeletal muscle has no pacemaker cells and shows no spontaneous activity.
(b) **False** Calcium is released from intracellular stores when it contracts.
(c) **False** The filaments do not shorten but slide together over one another.
(d) **True** There is greater overlap of the actin and myosin fibrils.
(e) **True** Moderate stretch increases contraction strength as in the heart.

235 (a) **False** EEG is no index of intelligence.
(b) **False** High-amplitude waves occur in deep sleep; the reverse is true of wakefulness.
(c) **True** The high-amplitude waves in sleep are of low frequency; the reverse is true of the low-amplitude waves in wakefulness.
(d) **False** EEG voltages are much smaller than ECG voltages.
(e) **True** Asymmetry is a sign of disease.

236 (a) **True** Excitation leaps from node to node across the myelinated segments.
(b) **False** The membrane does depolarize at the nodes.
(c) **True** Cooling slows sodium conductance at the nodes.
(d) **True** Theoretically, saltatory conduction becomes slower in nerve fibres with diameters less than 1 μm.
(e) **True** And to the distance between nodes.

237 Parasympathetic nerves:
   (a) Have opposite effects to sympathetic nerves on intestinal smooth muscle
   (b) Have opposite effects to sympathetic nerves on iris smooth muscle
   (c) Cause vasodilation in skeletal muscle during prolonged exercise
   (d) Cause sweat secretion in skin when body temperature rises
   (e) Have longer postganglionic than preganglionic fibres

238 Alpha (α) adrenoceptors:
   (a) Occur on myofilaments in smooth muscle cells
   (b) Are distinguishable from beta (β) receptors using electron microscopy
   (c) Can be stimulated by both adrenaline and noradrenaline
   (d) Are involved in the vasoconstrictor responses in skin to adrenaline
   (e) Are involved in the heart rate responses to noradrenaline

239 Primary neurones serving conscious muscle proprioception:
   (a) Conduct impulses at a similar rate to somatic motor neurones
   (b) Have their cell bodies in the ipsilateral posterior horn of the spinal cord
   (c) Use a different pathway from the primary neurones serving unconscious proprioception
   (d) Synapse with secondary neurones whose axons project up the ipsilateral posterior (dorsal) columns of the spinal cord
   (e) Synapse with neurones which cross the midline of the body in the brain stem

240 The α (alpha) rhythm of the electroencephalogram:
   (a) Disappears when the eyes are closed
   (b) Is an electrical potential with an amplitude around 1 mV
   (c) Has a frequency of 8–12 Hz
   (d) Has a lower frequency than the δ (delta) rhythm
   (e) Indicates that the subject is awake

241 Nerve fibres continue to conduct impulses when:
   (a) Extracellular sodium is replaced by potassium
   (b) Extracellular sodium is replaced by a non-diffusible cation
   (c) Temperature is lowered from 37 to 30°C
   (d) Temperature is lowered to below 0°C provided freezing does not occur
   (e) The sodium–potassium pump is inactivated

242 The primary sensory ending of a muscle spindle is stimulated by:
   (a) Shortening of an antagonist muscle
   (b) Relaxation of the muscle concerned when under load
   (c) Shortening of the extrafusal fibres
   (d) Stimulation of the gamma (γ) efferent fibres
   (e) Striking the appropriate tendon with a tendon hammer

237 (a) True Parasympathetic nerves stimulate, and sympathetic nerves inhibit, intestinal smooth muscle contractions.

(b) False Both contract iris smooth muscle; however, parasympathetics constrict the pupil by contracting circular muscle, sympathetics dilate it by contracting radial muscle.

(c) False Skeletal muscle has no parasympathetic nerve supply; local metabolites are responsible for the vasodilation.

(d) False Skin has no parasympathetic innervation; sympathetic cholinergic nerves are responsible for the increase in sweating when body temperature rises.

(e) False The reverse is the case.

238 (a) False They occur in the cell membranes.

(b) False They cannot be visualized; chemical evidence suggests their structure is similar.

(c) True Also by drugs such as ephidrine which has a similar structure.

(d) True Alpha receptors predominate in skin arterioles.

(e) False Cardiac adrenoceptors are predominantly beta receptors.

239 (a) True Both are fast fibres and reflexes involving them have a short reflex delay.

(b) False They are in the posterior root ganglion.

(c) True Unconscious proprioception impulses travel in the spino-cerebellar tracts.

(d) False The primary neurone axons pass up the ipsilateral posterior columns before synapsing with secondary neurones at the top of the spinal cord.

(e) True The secondary neurones cross in the medulla oblongata.

240 (a) False It is best seen when the subject's eyes are closed.

(b) False It is much smaller, at around 50 μV.

(c) True It is described in terms of frequency and amplitude.

(d) False The upper limit of the delta rhythm is 3.5 Hz.

(e) True It disappears with the onset of sleep.

241 (a) False This would depolarize the fibres completely.

(b) False Influx of cations is essential for depolarization.

(c) True However, conduction is slowed.

(d) False Nerve fibres stop conducting before tissue freezing occurs.

(e) True They continue to conduct until the electrochemical gradients created by the pump decline.

242 (a) True This stretches the agonist and its spindles.

(b) True This also stretches the spindles.

(c) False This reduces the stretch of the spindle.

(d) True Muscle contraction at each end of the spindle stretches the nuclear bag.

(e) True The knee jerk is initiated by striking the patellar tendon which stretches spindles in the quadriceps muscle.

243 In the spinal cord:
   (a) Pain impulse traffic may be modulated in the posterior horn
   (b) Autonomic motor neurones arise in the lateral horn
   (c) Gamma-aminobutyric acid may act as an excitatory neurotransmitter
   (d) Reflex centres are normally inhibited by descending impulses from supra-spinal centres
   (e) Postsynaptic excitation may be mediated by amino acid derivatives acting as neurotransmitters

244 In the cerebral cortex:
   (a) Neuronal connections are innate and immutable
   (b) Language and non-language skills are represented in different hemispheres
   (c) The areas concerned with emotional behaviour are concentrated in the frontal lobes
   (d) The cortical area devoted to sensation in the hand is larger than that for the trunk
   (e) Stimulation of the motor cortex causes contractions of individual muscles on the opposite side of the body

245 When a nerve cell membrane is depolarized by 5 mV:
   (a) Its permeability to sodium increases
   (b) Sodium ions move into the cell to cause further depolarization
   (c) Potassium ions move outwards down their electrochemical gradient
   (d) Chloride ions move inwards down their electrochemical gradient
   (e) An action potential is generated

246 Generalized sympathetic activity is characterized by:
   (a) Contraction of the radial muscle in the iris
   (b) Increased urinary excretion of catecholamines
   (c) Lipolysis in adipose tissue
   (d) Decreased conduction rate in the atrio-ventricular bundle
   (e) Relaxation of sphincteric smooth muscle in the alimentary tract

247 Stimulation of the vagus nerves causes:
   (a) A reduction in the strength of ventricular contraction
   (b) Secretion and vasodilation in the salivary glands
   (c) Mucous secretion from bronchial mucosal cells
   (d) The spleen to contract
   (e) The gall bladder to contract

243 (a) True In the substantia gelatinosa; endorphins may inhibit pain impulse transmission.

(b) True These are the preganglionic autonomic motor neurones.

(c) False It is an inhibitory transmitter thought to be responsible for IPSPs at postsynaptic membranes.

(d) False Loss of supraspinal facilitation in responsible for the areflexia in spinal shock.

(e) True Glutamate and aspartate are thought to be responsible for much of the excitatory transmission in the CNS.

244 (a) False The connections of brain neurones can be changed fairly rapidly to reflect new patterns of activity and sensory experience; this is referred to as neuronal plasticity.

(b) True The left hemisphere is dominant for speech in most people and the right is dominant for skills requiring appreciation of time and space relationships.

(c) False They are found mainly in the limbic cortex.

(d) True The cortical area given to a particular skin area is related to the richness of its sensory innervation, not its anatomic size.

(e) False Stimulation causes integrated movements not individual muscle contractions.

245 (a) True Depolarization opens sodium channels in the membrane.

(b) True Positively charged sodium ions move down their electrochemical gradient, making the inside of the fibre more positive.

(c) True The fall in membrane potential increases the outward electrochemical gradient for the positively charged potassium ions; their exit tends to stabilize the membrane potential.

(d) True The fall in membrane potential increases the inward electrochemical gradient for chloride; this again tends to restore the resting membrane potential.

(e) False Not unless it reaches the threshold for firing.

246 (a) True This dilates the pupil.

(b) True Derived from activity in the adrenal medulla and sympathetic nerve endings.

(c) True This provides fatty acids for energy production.

(d) False It increases and the PR interval shortens; higher heart rates are made possible.

(e) False Sympathetic activity inhibits most smooth muscle in gut but contracts the sphincters.

247 (a) False Ventricles have little vagal innervation.

(b) False The vagus does not supply parasympathetic nerves to the salivary glands.

(c) True It also contracts bronchial smooth muscle.

(d) False It is sympathetic activity that causes the spleen to contract.

(e) True This, together with relaxation of the sphincter of Oddi, drives bile into the duodenum.

248 An action potential in a nerve fibre:
- (a) Occurs when its membrane potential is hyperpolarized to a critical level
- (b) Is associated with a transient increase in membrane permeability to sodium
- (c) Is associated with a transient decrease in membrane permeability to potassium
- (d) Induces local electrical currents in adjacent segments of the fibre
- (e) Has an amplitude which varies directly with the strength of stimulus

249 Non-myelinated axons differ from myelinated axons in that they are:
- (a) Not sheathed in Schwann cells
- (b) Not capable of regeneration after section
- (c) Found only in the autonomic nervous system
- (d) Less excitable
- (e) Refractory for a longer period after excitation

250 Resting nerve cell membranes are more permeable to:
- (a) Organic anions than to Cl^- anions
- (b) K^+ ions than to Cl^- ions
- (c) Na^+ ions than to K^+ ions
- (d) Oxygen molecules than to glucose molecules
- (e) Water molecules than to H^+ ions

251 Acetylcholine:
- (a) Acts on the same type of receptor on postganglionic fibres in sympathetic and parasympathetic ganglia
- (b) Acts on the same type of receptor on target organs at cholinergic sympathetic and parasympathetic nerve terminals
- (c) Acts on the same type of receptor at autonomic ganglia and at somatic neuromuscular junctions
- (d) Acts as an excitatory transmitter in the basal ganglia
- (e) In blood is hydrolysed by the same cholinesterase as is found at neuromuscular junctions

252 Visceral smooth muscle differs from skeletal muscle in that:
- (a) It contracts when stretched
- (b) It is not paralysed when its motor nerve supply is cut
- (c) Its cells have unstable resting membrane potentials
- (d) It contains no actin or myosin
- (e) Excitation depends more on influx of extracellular calcium than on release of calcium from endoplasmic reticulum

248 (a) False It occurs when the membrane potential is reduced to its threshold for firing.

    (b) True This leads to rapid depolarization towards the sodium equilibrium potential.

    (c) False Permeability to K^+ increases and the resulting K^+ efflux contributes to membrane repolarization.

    (d) True This depolarizes the adjacent axon and leads to propagation of the impulse.

    (e) False Because of the 'all or none' law, impulse configuration is independent of stimulus strength.

249 (a) False Both types have Schwann cell sheaths.

    (b) False Following section, the central end of the axon buds and grows down the Schwann cell sheath until it reaches its target organ.

    (c) False Some sensory fibres serving pain and temperature are unmyelinated C fibres.

    (d) True Myelinated fibres have a lower threshold for stimulation.

    (e) True 2 ms compared with 0.5 ms in well-myelinated nerves; myelinated fibres can transmit impulses at higher frequencies than unmyelinated nerves.

250 (a) False Organic anions cannot cross the membrane readily.

    (b) False The permeability to Cl^- is about twice that to K^+.

    (c) False K^+ permeability is about 100 times that of Na^+.

    (d) True O_2, being a small, non-polar, lipophilic molecule, dissolves in the membrane lipid and crosses more easily than the larger glucose molecule.

    (e) True The bilipid layer is more permeable to water than to the charged hydrogen ion.

251 (a) True These are 'nicotinic' receptors and the action is blocked at both sites by drugs such as hexamethonium.

    (b) True These are 'muscarinic' receptors and the action is blocked at both sites by atropine.

    (c) False The receptors are different; curare blocks transmission at somatic neuromuscular junctions but not at ganglia.

    (d) True Anticholinergic drugs are sometimes useful in the treatment of muscle rigidity in Parkinsonism.

    (e) False The 'pseudocholinesterase' found in blood differs from the 'true cholinesterase' found near neuromuscular junctions.

252 (a) True The intrinsic 'myogenic' response in smooth muscle opposes stretch: skeletal muscle requires a nerve reflex arc for this type of response.

    (b) True It continues to contract due to local pacemakers.

    (c) True They show spontaneous depolarization between contractions.

    (d) False In smooth muscle, the actin and myosin filaments are less obvious on microscopy.

    (e) True Smooth muscle has a less well-developed sarcoplasmic reticulum.

253 An inhibitory postsynaptic potential:
   (a) May be recorded in a postganglionic sympathetic neurone
   (b) May be recorded in an anterior horn motor neurone
   (c) Does not exceed 1 mV in amplitude
   (d) Moves membrane potential towards the equilibrium potential for potassium
   (e) May summate in space and time with other excitatory and inhibitory potentials in the same neurone

254 A volley of impulses travelling in a presynaptic neurone causes:
   (a) An identical volley in the postsynaptic neurone
   (b) An increase in the permeability of the presynaptic nerve terminals to calcium
   (c) Vesicles in the nerve endings to fuse with the cell membrane and release their contents
   (d) The generation of at least one action potential in the postsynaptic neurone
   (e) Neurotransmitter to travel down the nerve axon

255 Pain receptors are:
   (a) Similar in structure to Pacinian corpuscles
   (b) Stimulated by a rise in the local K^+ concentration
   (c) Quick to adapt to a constant stimulus
   (d) More easily stimulated in injured tissue
   (e) Stimulated in the wall of the gut by agents which damage the tissues

256 A property shared by:
   (a) Skeletal and cardiac muscle is their striated microscopical appearance
   (b) Skeletal and multiunit smooth muscle is that they are paralysed when their motor nerves are cut
   (c) Cardiac and visceral smooth muscle is their spontaneous activity when denervated
   (d) Skeletal and cardiac ventricular muscle is their stable resting membrane potential
   (e) All varieties of muscle is that contraction strength is related to their initial length

253 (a) **False** No such potentials have been recorded here.

   (b) **True** The hyperpolarization of the anterior horn cell reduces the likelihood of the cell firing action potentials by moving the membrane potential further from the firing threshold.

   (c) **False** Its amplitude is about 5 mV and its duration about 5 ms.

   (d) **True** It may be produced by increased permeability to potassium or to chloride ions.

   (e) **True** In this way the postsynaptic cell integrates the various signals it receives.

254 (a) **False** The synapse may amplify or attenuate the signal.

   (b) **True** The uptake of Ca^{2+} by the nerve ending facilitates release of transmitter.

   (c) **True** The neurotransmitter contained in the vesicles is released by exocytosis.

   (d) **False** The impulses may be inhibitory; even if they are excitatory, the postsynaptic neurone may be strongly inhibited by inputs from other preganglionic neurones.

   (e) **True** Neurotransmitter is thought to be synthesized and packaged in the Golgi apparatus in the neurone cell body before travelling down the axon to the nerve terminals.

255 (a) **False** They are bare nerve endings.

   (b) **True** K^+ salts cause pain when applied to the base of an ulcer.

   (c) **False** Pain receptors adapt very slowly; this protects the tissues against damage.

   (d) **True** The sensitivity of pain receptors is increased by local tissue injury (hyperalgesia).

   (e) **False** Damage to the gut wall by stimuli such as cutting or burning is painless.

256 (a) **True** Both have highly organized actin and myosin filaments.

   (b) **True** The iris is an example of multiunit smooth muscle.

   (c) **True** Isolated hearts and gut segments show spontaneous activity in the organ bath.

   (d) **True** Only the pacemaker cells in the heart have unstable membrane potentials.

   (e) **True** The Frank–Starling relationship describes this with respect to cardiac muscle.

257 The equilibrium potential (E) for:
(a) An ion species is the membrane potential observed when its concentrations on each side of the membrane are in equilibrium
(b) Na^+ is about −50 mV in squid axon
(c) An ion species depends on the ratio of the concentrations of the ion outside (I_o) and inside (I_i) the cell
(d) An ion species is the potential that the membrane potential would approach if it became freely permeable to that ion
(e) An ion species would be zero if the concentrations of the ion on each side of the membrane were equal

258 Histological and physiological study of skeletal muscle shows that the:
(a) Distance between two Z lines remains constant during contraction
(b) Width of the anisotropic (A) band is constant during contraction
(c) Tension developed is maximal when actin and myosin molecules just fail to overlap
(d) Stimulus needed to cause contraction is minimal when applied at the Z line
(e) The T system of transverse tubules opens into the terminal cisterns of the sarcoplasmic reticulum

259 Rapid eye movement (REM) sleep differs from non-REM sleep in that:
(a) The EEG shows waves of higher frequency
(b) Muscle tone is higher
(c) Heart rate and respiration are more regular
(d) Secretion of growth hormone is increased
(e) Dreaming is more common

260 Hemisection of the spinal cord at C7 on the right side would cause:
(a) Greater loss of pain sensation in the right foot than in the left foot
(b) Greater loss of motor power in the right leg than in the left leg
(c) Greater loss of conscious proprioception in the right than in the left leg
(d) Respiratory failure
(e) Loss of the micturition reflex

261 The ankle jerk reflex is exaggerated:
(a) When the arm muscles are voluntarily contracted
(b) Immediately after complete spinal cord transection at the cervical level
(c) On the left side some months after damage to the tracts in the right internal capsule
(d) In extrapyramidal system disorders such as Parkinsonism
(e) When cerebellar function is lost

257 (a) **False**   It is the membrane potential that would be required to balance the concentration gradient for that ion.

(b) **False**   A positive membrane potential of about +65 mV is required to balance the Na^+ concentration gradient.

(c) **True**   It can be calculated using the Nernst equation:
$E = 61.5 \log_{10} [I_o]/[I_i]$ valence^{-1} at 37°C.

(d) **True**   Membrane potential approaches +65 mV when freely permeable to Na^+.

(e) **True**   From the Nernst equation ($\log_{10} 1 = 0$).

258 (a) **False**   This is sarcomere length and it shortens with contraction.

(b) **True**   Its width is the length of the myosin molecule.

(c) **False**   It is maximal when action and myosin overlap maximally and when adjoining actin molecules just fail to overlap.

(d) **True**   This is where the transverse tubules penetrate the muscle fibre.

(e) **False**   They are adjacent but are not directly connected.

259 (a) **True**   Wave amplitude is reduced also.

(b) **False**   Surprisingly, muscle tone is low.

(c) **False**   They are much more irregular.

(d) **False**   It is inhibited during REM sleep.

(e) **True**   Subjects woken at the end of REM sleep give vivid accounts of dreams.

260 (a) **False**   The left would be more affected; pain fibres cross the midline shortly after entering the cord.

(b) **True**   The main motor tract to the right leg would be severed.

(c) **True**   The fibres serving proprioception cross at the top of the spinal cord.

(d) **False**   Diaphragmatic and some intercostal activity would remain intact.

(e) **False**   The reflex centres for micturition are in the sacral cord.

261 (a) **False**   It is not exaggerated but contraction of arm muscles reinforces the reflex.

(b) **False**   Immediately after cord transection, reflexes below the level of the lesion are lost in the stage of 'spinal shock'.

(c) **True**   Muscle tone increases following contralateral internal capsular lesions.

(d) **False**   Though muscle tone is increased, tendon jerks are not exaggerated.

(e) **False**   Muscle tone is low following cerebellar damage and the knee jerk 'pendular'.

262 Delta (δ) wave activity in the electroencephalogram:
(a) Is low in frequency and amplitude
(b) Suggests that the patient is alert and concentrating
(c) When unilateral suggests a brain abnormality
(d) Is a feature of petit mal epilepsy
(e) Is more common in children than in adults while they are awake

263 Increased intracranial pressure may cause:
(a) Cranial enlargement in children
(b) Squinting and loss of smell sensation in children
(c) Cupping of the optic disc
(d) An increase in cerebral blood flow
(e) Arterial hypertension

264 Pain receptors in the gut and urinary tract may be stimulated by:
(a) Cutting through their wall with a sharp scalpel
(b) Distension
(c) Inflammation of the wall
(d) Acid fluid
(e) Vigorous rhythmic contractions behind an obstruction

265 Signs of brain-stem death include:
(a) Unconsciousness
(b) Loss of pupillary reaction to light
(c) Loss of tendon jerks in the arms and legs
(d) Loss of respiratory response to CO_2 in the absence of hypoxia
(e) Nystagmus in response to cold water in the external auditory canal

266 Spasm of digital vessels in the hands in response to cold (Raynaud's phenomenon) may be relieved by:
(a) Cutting sympathetic nerves to the hand
(b) Stimulating parasympathetic nerves to the hand
(c) Alpha (α) adrenoceptor blockade
(d) Beta (β) adrenoceptor agonists
(e) Wearing gloves

267 Muscle tone is reduced by:
(a) Curare-like drugs
(b) Lower motor neurone lesions
(c) Upper motor neurone lesions
(d) Cerebellar lesions
(e) Gamma efferent impulses to muscle spindles

262 **(a) False** Delta waves have low frequency and large amplitude.
(b) False It is associated with deep sleep.
(c) True Delta waves may indicate brain damage as well as deep sleep.
(d) False This produces a characteristic spike and wave pattern in the EEG.
(e) True Delta wave activity is seen occasionally in the EEGs of normal awake children.

263 **(a) True** Hydrocephaly occurs because the cranial bone sutures have not closed.
(b) True The cranial deformity in hydrocephalus may damage cranial nerves.
(c) False It causes papilloedema – bulging in the opposite direction.
(d) False Vessel compression may reduce cerebral flow severely.
(e) True This reflex response helps to maintain cerebral blood flow.

264 **(a) False** The intestine may be cut painlessly during operations under local anaesthesia.
(b) True Stretch is an adequate stimulus for these receptors.
(c) True Chemicals released in inflammation lower the pain threshold (hyperalgesia).
(d) True As with a peptic ulcer.
(e) True This is the cause of the intermittent pain known as colic.

265 **(a) False** Consciousness is a function of the cerebral cortical neurones.
(b) True The coordinating centres for pupillary reflexes are in the brain stem.
(c) False The reflex centres for tendon jerks are in the spinal cord.
(d) True Respiratory reflexes arc coordinated in the brain stem.
(e) False Nystagmus is the normal reflex response of brain-stem centres to this stimulus.

266 **(a) True** Sympathetic nerves to digital vessels are tonic vasoconstrictor nerves.
(b) False The hand does not have a parasympathetic nerve supply.
(c) True Alpha receptors mediate sympathetic vasoconstriction.
(d) False Digital vessels have few beta receptors.
(e) True Raynaud's phenomenon is triggered by cold; wearing gloves may prevent cooling to the threshold temperature.

267 **(a) True** These paralyse muscle by blocking transmission at neuromuscular junctions.
(b) True Lower motor neurone lesions also paralyse skeletal muscle.
(c) False Loss of supraspinal influences results in spasticity of the affected muscles.
(d) True The cerebellum helps to maintain normal muscle tone.
(e) False These increase spindle sensitivity to stretch and hence increase muscle tone.

268 Effective treatments for severely raised intracranial pressure include:
   (a) Removal of cerebrospinal fluid by lumbar puncture
   (b) Reduction of extracellular fluid by diuretics
   (c) Creating an anastomosis between a cerebral ventricle and a neck vein
   (d) Creating an anastomosis between intracranial subdural space and the peritoneal cavity
   (e) Placing the patient recumbent with the legs raised

269 Disease of the extrapyramidal motor system in Parkinsonism causes:
   (a) Tremor which is more obvious when the patient is performing skilled movements
   (b) Muscle paralysis
   (c) Increased muscle tone throughout the range of passive movement
   (d) Increased involuntary facial movements during speech
   (e) An unusual gait with small fast regular steps

270 Lower motor neurone disease:
   (a) Causes loss of voluntary movements but not of reflex movements
   (b) Is a later stage of upper motor neurone disease
   (c) Causes eventual wasting of the muscles concerned
   (d) Does not affect ventilation of the lungs
   (e) Is associated with involuntary twitchings of small fasciculi in the affected muscles

271 Blockade of α (alpha) adrenoceptors is likely to cause a reduction in:
   (a) Sweat production
   (b) Bronchus diameter
   (c) Gastrointestinal motility
   (d) Total peripheral resistance
   (e) Heart rate

272 Bulging of the optic disc into the vitreous humour (papilloedema) is caused by:
   (a) Raised intraocular pressure (glaucoma)
   (b) Blockage of absorption of the aqueous humour
   (c) A rise in intracranial pressure
   (d) Inflammation of the optic nerve
   (e) Interference with the venous drainage of the eye

273 Atropine (which blocks muscarinic receptors) causes:
   (a) Paralysis of accommodation for near vision in the eye
   (b) Constriction of the pupil
   (c) Constriction of the bronchi
   (d) Diarrhoea
   (e) Difficulty with micturition

268 (a) **False** This may cause fatal compression of the brain stem if the medulla is forced (coned) into the foramen magnum.
(b) **True** This reduces formation of cerebrospinal fluid.
(c) **True** This facilitates drainage of cerebrospinal fluid.
(d) **False** CSF is in the subarachnoid space.
(e) **False** This would increase intracranial pressure due to gravitational effects and impair CSF drainage by raising pressure in the venous sinuses.

269 (a) **False** Parkinsonism causes tremor which is more obvious when the patient is at rest.
(b) **False** Paralysis is not a feature of the condition.
(c) **True** This is 'cogwheel' or 'lead pipe' rigidity.
(d) **False** The face is mask-like; there is poverty of facial movements.
(e) **True** The body's centre of gravity is shifted forward.

270 (a) **False** The muscles are paralysed and not capable of voluntary or reflex movements.
(b) **False** It is totally independent of upper motor neurone disease.
(c) **True** It leads to 'disuse atrophy'.
(d) **False** Ventilation will be affected if the phrenic and intercostal nerves are involved.
(e) **True** 'Fasciculation' is due to denervation hypersensitivity; muscles become ultrasensitive to small amounts of acetylcholine released from the degenerating nerve terminals.

271 (a) **False** Sweating is mediated by cholinergic nerves.
(b) **False** Bronchial smooth muscle has few alpha receptors; beta blockade tends to cause bronchoconstriction.
(c) **False** Stimulation of alpha or beta adrenoceptors inhibits the gut.
(d) **True** Alpha receptors mediate vasoconstriction.
(e) **False** The heart has few alpha receptors; beta receptor blockade slows the heart.

272 (a) **False** Glaucoma causes depression ('cupping') of the optic disc.
(b) **False** This is a cause of glaucoma.
(c) **True** Papilloedema is an important sign of raised intracranial pressure.
(d) **True** Optic neuritis causes swelling of the optic disc.
(e) **True** This causes local oedema involving the disc.

273 (a) **True** By paralysing the ciliary muscles.
(b) **False** It causes pupillary dilation by paralysing circular muscle in the iris.
(c) **False** It tends to dilate the bronchi by blocking vagal effects.
(d) **False** It causes constipation by suppressing bowel activity.
(e) **True** Micturition depends on cholinergic nerves.

274 Visceral pain:
- (a) Is poorly localized compared with pain arising in skin
- (b) Is often felt in the mid line
- (c) May cause reflex contraction of overlying skeletal muscle
- (d) May cause reflex vomiting
- (e) May cause reflex changes in arterial blood pressure

275 Posterior column damage in the spinal cord may impair:
- (a) Vibration sense
- (b) Pain sensation
- (c) The flexor plantar response to stimulation of the sole
- (d) Touch sensation
- (e) The ability to stand steadily with the eyes closed

276 Aphasia:
- (a) Is an impairment of language skills without motor paralysis, loss of hearing or vision
- (b) Implies impairment of consciousness
- (c) Is called motor aphasia if the patient understands what the speech sounds and symbols mean but lacks the higher motor skills needed to express them
- (d) Is called sensory aphasia if the patient does not understand the meaning of the words he hears, sees and uses
- (e) Usually results from right-sided cortical damage

277 Blockade of parasympathetic activity causes a reduction in:
- (a) Sweat production
- (b) Resting heart rate
- (c) The strength of skeletal muscle contraction
- (d) Salivation
- (e) Intestinal motility

278 Sensory disturbance consisting of:
- (a) Pain, sensory loss and paraesthesiae in one leg suggests a spinal cord lesion
- (b) Loss of pain, temperature but not touch sensation in the arms suggests a spinal cord lesion
- (c) Loss of two-point discrimination but not touch sensation suggests a lesion in the thalamus
- (d) Loss of all sensations on the left side suggests a right internal capsule lesion
- (e) Loss of all sensations in a skin region suggests a peripheral nerve or posterior root lesion

274 (a) True This is characteristic of visceral pain.
   (b) True However, pain from some viscera such as ureters and bile ducts is lateralized.
   (c) True In the abdomen this is seen as 'guarding'.
   (d) True Visceral pain is often associated with autonomic side-effects.
   (e) True Another example of autonomic side-effects, pressure may increase or decrease.

275 (a) True The fibres carrying this sensation travel in the posterior columns.
   (b) False The fibres carrying pain sensation travel in the spinothalamic tracts.
   (c) False The reflex does not depend on posterior column fibres.
   (d) True The fibres responsible for fine touch sensation travel in the posterior columns.
   (e) True Proprioceptive information is important in maintaining balance and it is carried in the posterior columns.

276 (a) True It is a cortical dysfunction.
   (b) False Level of consciousness is an independent entity.
   (c) True It may start with an inability to say the names of familiar objects but end up with loss of virtually all verbal communication skills.
   (d) True Patients with sensory aphasia tend to talk rubbish since they are unaware of the errors in their use of language.
   (e) False Language skills are carried in the left hemisphere in right-handed and some left-handed people.

277 (a) False Sweat glands are innervated by sympathetic cholinergic nerves.
   (b) False Resting heart rate rises due to blockade of vagal tone.
   (c) False Parasympathetic nerves are not involved in skeletal muscle activity.
   (d) True Dryness of the mouth results from blockade of salivary secretion.
   (e) True Parasympathetic nerves are motor to intestinal smooth muscle.

278 (a) False Pain is uncommon with spinal cord lesions; the symptoms suggest irritation of a sensory root or peripheral nerve.
   (b) True The fibres carrying these sensations run in separate tracts in the spinal cord.
   (c) False It suggests a parietal cortex lesion where such sensory discriminations are made.
   (d) False It suggests right-sided brain-stem damage since pain is appreciated at subcortical level.
   (e) True Only peripheral nerves and posterior roots carry all modalities of sensation together from a circumscribed skin area.

279 Damage to the cerebral cortex may cause loss of:
   (a) Pain sensation on the opposite side of the body
   (b) Reflex thermoregulatory activity
   (c) Skilled movements in the absence of paralysis
   (d) Ability to identify an object by its tactile characteristics
   (e) Vision in one eye only

280 In the hemiplegia following a right-sided cerebrovascular accident (stroke):
   (a) Left-sided muscle weakness is evident
   (b) Muscles in the left side of the body are unable to contract
   (c) Muscles which act on both sides of the body, such as respiratory muscles, tend to be spared
   (d) Skilled movements are better preserved than unskilled movements
   (e) Speech movements are better preserved than swallowing movements

281 Characteristic features of cerebellar disease include loss of:
   (a) Muscle tone
   (b) Muscle strength
   (c) Conscious muscle-joint sense
   (d) Ability to make precise muscle movements
   (e) Ability to fix the gaze steadily on an object

282 Long-term consequences of transection of the spinal cord in the lower cervical region include:
   (a) Loss of thermoregulatory sweat production in the legs
   (b) Severe flexor spasms when the skin of the legs is stimulated
   (c) Paralysis of bladder muscle
   (d) Inability to regulate sympathetic tone in leg blood vessels in response to baroreceptor stimulation
   (e) Inability to erect the penis and ejaculate semen

283 Blockade of β (beta) adrenoceptors is likely to cause:
   (a) Decreased intestinal motility
   (b) Worsening of the condition in patients with bronchial asthma
   (c) Worsening of the condition in patients in cardiac failure
   (d) Inability to increase heart rate during exercise in patients with transplanted hearts
   (e) Inability to increase blood flow to exercising muscles

279 **(a)** **False** The cerebral cortex is not involved in the conscious appreciation of pain.
   (b) **False** Reflex thermoregulation is organized in the hypothalamus.
   (c) **True** Such loss of skilled movements is called 'apraxia'.
   (d) **True** This 'asteriognosia' results from parietal cortex damage.
   (e) **False** Damage to the visual cortex causes loss of part of the visual field in both eyes.

280 **(a)** **True** Voluntary movements such as making a hand grip are abnormally weak.
   (b) **False** Voluntary control is lost but the muscles can contract in reflex, synergistic and other involuntary movements.
   (c) **True** Perhaps because they have bilateral cortical representation.
   (d) **False** The most affected movements are those requiring high levels of cortical control.
   (e) **False** The vocal cord movements required for speech are more highly skilled that those required for swallowing.

281 **(a)** **True** Muscle tone depends in part on the integrity of the cerebellum.
   (b) **False** Muscle strength does not depend on the cerebellum.
   (c) **False** Cerebellar activity depends on unconscious, not conscious, proprioception.
   (d) **True** Loss of muscular coordination causes 'intention tremor'.
   (e) **True** The jerky eye movement in cerebellar disease when the gaze is fixed on an object is called 'nystagmus'.

282 **(a)** **True** Thermoregulatory control is coordinated through the sympathetic nervous system by centres in the hypothalamus.
   (b) **True** Due to exaggeration of the spinal withdrawal reflex.
   (c) **False** Both micturition and defaecation can occur reflexly (their reflex centres are in the sacral cord) but are no longer under voluntary control.
   (d) **True** This impairs the control of arterial blood pressure, especially in the erect position.
   (e) **False** These are spinal reflexes with centres in the lumbosacral region of the spinal cord.

283 **(a)** **False** Alpha and beta adrenoceptor simulation in gut inhibits smooth muscle.
   (b) **True** By blocking the bronchodilatory action of adrenaline and noradrenaline.
   (c) **True** By several mechanisms including blocking the sympathetic drive to the heart.
   (d) **True** By blocking the chronotropic effect of circulating catecholamines.
   (e) **False** Increased flow in exercise depends on local metabolites, not beta adrenoceptors.

284 Cutaneous pain:
  (a) Can be caused by overstimulation of touch receptors
  (b) Can be caused by excitation of receptors by chemicals released in injured tissue
  (c) Can be elicited more readily if the tissue has been injured recently
  (d) Receptors adapt to stimulation more quickly than touch receptors
  (e) Transmission at spinal cord level is facilitated by the opening of potassium channels in the postsynaptic membrane

285 Headache can be produced by:
  (a) Dilation of intracranial blood vessels
  (b) Constriction of extracranial blood vessels
  (c) Meningeal irritation
  (d) Blood in the cerebrospinal fluid
  (e) Loss of cerebrospinal fluid following lumbar puncture

286 Depressed brain function causes:
  (a) Disorders of function which are independent of the cause of the depression
  (b) Loss of unskilled movements before loss of skilled movements
  (c) Restlessness and delirium before stupor and coma
  (d) A progressive fall in the amplitude of waves in the EEG
  (e) Depression of ventilation before loss of consciousness

287 Loss of pain sensation in the:
  (a) Feet may lead to skin ulceration
  (b) Knee may lead to joint damage
  (c) Ears and fingers usually precedes frostbite
  (d) Leg follows surgical division of the spinocerebellar tracts in the spinal cord
  (e) Face follows surgical division of the facial nerve

288 Intracranial pressure tends to rise when:
  (a) Cerebral venous pressure rises
  (b) Forced expiration is made against a closed glottis
  (c) There is a bout of coughing
  (d) Cerebral blood flow increases
  (e) Arterial P_{CO2} falls below normal

284 (a) False Pain is due to excitation of specific pain receptors.
(b) True Such substances include histamine, bradykinin, 5-hydroxytryptamine, prostaglandins, NO and substance P.
(c) True Chemicals released in the injured tissue reduce the stimulation threshold to cause hyperalgesia.
(d) False Pain receptors adapt much more slowly than touch receptors.
(e) False This would inhibit synaptic transmission and may explain how enkephalin and endorphin modulate transmission of pain impulses in the substantia gelatinosa.

285 (a) True Probably due to pain receptors in vessel walls stimulated by stretch.
(b) False Dilation of these extracranial vessels occurs in patients with throbbing migraine-type headaches that are relieved by certain vasoconstrictor agents.
(c) True This causes the severe pain and neck stiffness in early meningitis.
(d) True Surprisingly, blood also irritates the meninges to cause headache and neck stiffness.
(e) True The brain tends to sag in the cranium and pull on pain receptors in the meninges.

286 (a) True Alcohol, lack of oxygen, low blood sugar and hypothermia all impair cellular function in the brain and lead to a recognizable syndrome of neurological effects.
(b) False Higher critical functions depending on cortical cells are first to go, e.g. driving ability.
(c) True Coma implies that cellular depression has reached the less sensitive cells in the brain stem mediating consciousness.
(d) False Amplitude increases initially to give delta waves in light coma before falling later.
(e) False This brain-stem function is preserved until the later stages of deep coma.

287 (a) True Probably because of failure to protect the foot and letting minor injury persist without attention.
(b) True Again due to loss of the protective effects of pain.
(c) True Nerve fibres stop conducting near freezing point so there is no further warning when freezing causes tissue damage.
(d) False Pain sensation is blocked by surgical division of the spinothalamic tracts.
(e) False It is lost following division of the sensory root of the trigeminal nerve.

288 (a) True This distends intracranial veins and hence raises pressure within the cranium.
(b) True This (Valsalva) manoeuvre raises cerebral venous pressure.
(c) True This also raises cerebral venous pressure.
(d) True The vasodilation raises the volume of blood within the cranium.
(e) False This constricts cerebral vessels and reduces cerebral blood volume.

289 Severe pain may lead to:
- (a) A fall in blood pressure due to a fall in vascular resistance in skeletal muscle
- (b) A fall in heart rate due to an increase in cardiac vagal tone
- (c) Vomiting through a reflex centre in the brain stem
- (d) Profuse sweating due to activation of sympathetic nerves
- (e) Suppression of cortisol secretion

289 **(a)** **True**   This lowers total peripheral resistance.

    (b) **True**   Consciousness may be lost if this is combined with muscle vasodilation in the 'vasovagal' syndrome.

    (c) **True**   The vomiting centre is in the medulla oblongata.

    (d) **True**   Sympathetic cholinergic nerves are motor to sweat glands.

    (e) **False**   Severe pain increases cortisol secretion via the hypothalamus as part of the general response to trauma.

6 SPECIAL SENSES

290 The fovea centralis:
  (a) Lies where the visual axis impinges on the retina
  (b) Is not crossed by any major blood vessels
  (c) Is the thickest part of the retina
  (d) Offers higher visual acuity than other parts of the retina
  (e) Lies on the temporal side of the optic disc

291 Endolymph:
  (a) Is found within the membranous labyrinth
  (b) Has a potassium concentration close to that of extracellular fluid
  (c) Bathes the hair cells of the inner ear
  (d) Is electrically negative with respect to perilymph
  (e) Inertia is a factor in the stimulation of receptors in the semicircular canals during rotatory acceleration

292 Olfactory cells:
  (a) Are epithelial cells which synapse with olfactory nerves
  (b) Generate impulses when stimulated which are relayed in the thalamus
  (c) Are chemoreceptors
  (d) Show little adaptation
  (e) Are more important than taste in appreciating the flavour of food

293 Adaptation for vision in poor light is:
  (a) Complete after 2–3 min
  (b) Due to pigment regeneration but not iris changes
  (c) Due to regeneration of rod but not cone pigments
  (d) Faster if red goggles are worn before entering the dark environment
  (e) More effective for central than for peripheral vision

294 The basilar membrane of the cochlea vibrates:
  (a) At the same frequency as the applied sound
  (b) With an amplitude which is proportional to the frequency of the applied sound
  (c) With an amplitude which is proportional to the loudness of the applied sound
  (d) Along more of its length when the applied sound has a high rather than a low frequency
  (e) Mainly at the base of the cochlea for the sound frequencies commonly used in speech

295 The cones in the retina differ from rods in that they are more:
  (a) Numerous
  (b) Concerned with colour vision
  (c) Sensitive to light
  (d) Concerned with high visual acuity
  (e) Affected by vitamin A deficiency

290 **(a) True**   It detects objects in the centre of the field of vision.
   (b) True   There are no superficial structures to affect impinging light rays.
   (c) False It is relatively thin due to the absence of superficial layers.
   (d) True   The above factors contribute to this.
   (e) True   It is marked by yellow pigment.

291 **(a) True**   Perilymph surrounds the membranous labyrinth.
   (b) False It is similar to that of intracellular fluid.
   (c) True   It bathes auditory and vestibular hair cells.
   (d) False It is positive, of the order of +80 mV.
   (e) True   The inertia causes endolymph movements to lag behind those of the membranous labyrinth and displace the hairs of the hair cells.

292 **(a) False** They are modified nerve cells in the nasal epithelium.
   (b) False Unlike other sensory inputs, olfactory impulses are not relayed in the thalamus.
   (c) True   They recognize certain molecular structures.
   (d) False It is the newcomer who notices the smell in the room.
   (e) True   In their absence, food loses much of its flavour.

293 **(a) False** It takes about 20 min.
   (b) False Dilation of the pupil also contributes.
   (c) False Regeneration of cone pigments plays a part.
   (d) True   Red light does not bleach rhodopsin.
   (e) False Peripheral vision adapts better to dark conditions.

294 **(a) True**   Harmonics are also faithfully reproduced.
   (b) False Frequency and amplitude need not be related.
   (c) True   Hence very loud sounds can damage the basilar membrane.
   (d) False Low-frequency vibrations travel further up the cochlea.
   (e) True   Speech frequencies (about 1000–3000 Hz) cause maximum vibration in this region.

295 **(a) False** There are about 6 million cones compared with 120 million rods.
   (b) True   Rods alone give achromatic vision.
   (c) False The rods are much more sensitive; their pigment is bleached in bright light.
   (d) True   Acuity is highest with foveal (cone) vision.
   (e) False Vitamin A is essential for rhodopsin synthesis in rod vision only.

296 Increasing the salt concentration applied to a 'salt' taste bud increases:
    (a) Its sensitivity to salt
    (b) The amplitude of its generator potentials
    (c) The amplitude of the action potentials generated
    (d) Impulse traffic to the thalamus
    (e) Impulse traffic up the ascending reticular formation

297 On entering a darkened room, the:
    (a) Threshold light intensity for the eye starts to rise
    (b) First phase of retinal adaptation is mainly in the cones
    (c) Final phase of retinal adaptation is mainly in the rods
    (d) Adaptation is slower if a long period was spent in bright light beforehand
    (e) Time course for pupillary dilation is similar to that for dark adaptation

298 Dilation of the pupil increases the:
    (a) Amount of light entering the eye
    (b) Refractive power of the eye
    (c) Spherical aberration
    (d) Depth of focus
    (e) Field of vision

299 The olfactory system can detect:
    (a) 20–40 distinct odours
    (b) Differences in odour between isomers of the same substance
    (c) The direction from which an odour comes
    (d) Small differences in the concentration of the substance responsible for the odour
    (e) Odours better in old than in young people

300 During accommodation for near vision:
    (a) More light enters the eye
    (b) The curvature of the cornea increases
    (c) Chromatic and spherical aberration is decreased
    (d) The depth of focus increases
    (e) The visual axes of the two eyes converge

301 Visual acuity is:
    (a) A measure of the sensitivity of the retina to light
    (b) Greater in a person with 6/12 (0.5) vision than in one with 6/9 (0.75)
    (c) Greater using central than using peripheral vision
    (d) Greater using one eye than using both eyes
    (e) Greater in normal than in colour-blind people

296 (a) False It decreases; taste receptors adapt to stimuli applied to them.
(b) True Stronger stimuli lead to generator potentials of greater amplitude.
(c) False There is an increase in impulse frequency, not amplitude.
(d) True All but olfactory impulses are relayed in the thalamus.
(e) True All sensory inputs send impulses via collaterals to this system.

297 (a) False It begins to fall in the process of dark adaptation.
(b) True The initial adaptation is due more to cone adaptation
(c) True Rods are slower to adapt but show more profound adaptation.
(d) True A previous long exposure would bleach most of the rhodopsin so more time would be needed to resynthesize it.
(e) False The pupil dilates almost immediately in the dark.

298 (a) True This allows rapid adaptation for vision in poor light.
(b) False The iris has nothing to do with the refractive power of the eye.
(c) True By permitting light to pass through peripheral, and less perfect, parts of the lens.
(d) False As with a camera, a wide aperture tends to shorten the depth of focus.
(e) False Narrowing the aperture in a camera does not result in a smaller picture.

299 (a) False It is thought that humans can differentiate between 2000 and 4000 different odours.
(b) True This may affect molecular configuration markedly.
(c) True Probably due to the different time of arrival of the odour at the two nostrils.
(d) False Though very low concentrations of odorous substances can be detected, differences in concentration of more than 30% are needed to detect a difference in intensity.
(e) False Olfaction ability falls with age.

300 (a) False Less light enters the eye as the pupils constrict.
(b) False It does not change; the lens ligaments relax, allowing it to become more spherical.
(c) True Due to pupillary constriction. Most spherical and chromatic aberration occurs at the periphery of the lens.
(d) True Due mainly to the improvement in optical characteristic mentioned in (c) above.
(e) True To allow the object to focus on the two foveae.

301 (a) False It is a measure of the ability to distinguish between (resolve) two points.
(b) False Acuity is expressed as the ratio of someone's reading distance compared with average normal.
(c) True Visual acuity is maximal at the fovea.
(d) False The two images reinforce each other.
(e) False Visual acuity does not depend on distinguishing colours.

302 The tympanic membrane:
  (a) Modifies the frequencies of sound waves impinging on the ear
  (b) Stops vibrating almost immediately after the sound stops
  (c) Bulges outwards when the pharyngotympanic tube is blocked
  (d) Transmits sound more effectively when the small muscles of the middle ear are contracted
  (e) Transmits sound more than 80% less efficiently when the membrane is perforated

303 In the refracting system of the eye:
  (a) The cornea causes more refraction than the lens
  (b) More refraction occurs at the inner surface of the cornea than at the outer surface
  (c) The lens, by becoming more convex, can double the total refractive power of the eye
  (d) The back surface of the lens contributes more to accommodation than the front
  (e) Ageing reduces the maximum refractive power of the eye

304 The hair cells in the semicircular canals are stimulated by:
  (a) Movement of perilymph
  (b) Linear acceleration
  (c) Rotation at constant velocity
  (d) Gravity
  (e) Movement of endolymph relative to hair cells

305 When light is shone into one eye, the pupil:
  (a) Constricts even though the optic nerve has been cut
  (b) Responds due to sympathetic nerve activity
  (c) Does not respond if autonomic cholinergic nerves are blocked
  (d) In that eye constricts and the opposite pupil dilates
  (e) Does not respond if there is brain-stem death

306 Light from an object to the right of the visual axis:
  (a) Impinges on the retina in the right eye to the right of the fovea
  (b) Impinges on the retina in the left eye to the left of the fovea
  (c) Generates impulses which travel in the right optic tract
  (d) Generates impulses which produce conscious sensation in the frontal lobe eye fields
  (e) Forms an inverted image on the retina

307 The tympanic membranes
  (a) Bulge inwards during descent in an unpressurized airplane
  (b) Have an area about twice that of the oval window
  (c) Prevent sound waves from reaching the oval and round windows at the same time
  (d) Transmit only 10% of applied sound energy to the cochlea for sound waves of 1000 Hz
  (e) Transmit sounds in the 500–5000 Hz frequency range with the least loss of energy

302 (a) **False** It faithfully reproduces the frequencies.
   (b) **True** It is very nearly critically damped.
   (c) **False** It bulges inwards as middle ear air is absorbed.
   (d) **False** Reflex contraction of these muscles protects by damping vibration transmission.
   (e) **False** Small perforations cause about 5 dB loss; complete destruction about 50.

303 (a) **True** This is because of the large change in refractive index from air to cornea.
   (b) **False** The outer interface is with air; cornea and aqueous have similar refractive indices.
   (c) **False** It can only increase total refractive power by about 15–20%.
   (d) **False** During accommodation, the front of the lens bulges more than the back.
   (e) **True** As the lens stiffens, ability to increase convexity when ciliary muscles contract is lost.

304 (a) **False** This is not in contact with the hair cells.
   (b) **False** Their adequate stimulus is rotary acceleration or deceleration.
   (c) **False** A blindfold person is unaware of any sensation when rotated at constant velocity, e.g. as on the earth!
   (d) **False** Gravity produces linear acceleration.
   (e) **True** This is the end-effect of the effective stimuli, angular acceleration and deceleration.

305 (a) **False** The optic nerve is an essential part of the reflex pathway for the light reflex.
   (b) **False** It still constricts after the sympathetic nerves are cut.
   (c) **True** Atropine blocks this parasympathetic action.
   (d) **False** Both pupils constrict consensually.
   (e) **True** This is a sign of brain-stem death.

306 (a) **False** It impinges to the left of the fovea.
   (b) **True** The retina of both eyes is stimulated in corresponding or homonymous areas.
   (c) **False** They travel in the left optic tract to the left occipital cortex.
   (d) **False** They enter consciousness in the left occipital cortex; frontal eye fields are concerned with eye movements.
   (e) **True** As in a camera.

307 (a) **True** Cabin pressure rises above middle ear pressure.
   (b) **False** Their area is about 20 times as great.
   (c) **True** This 'round window protection' prevents damping of vibrations in the inner ear.
   (d) **False** At this frequency, over half of the sound energy is transmitted.
   (e) **True** Impedency matching and auditory acuity are greatest at such frequencies which correspond to normal speech.

308 Utricles:
   (a) Are gravity receptors
   (b) Contain calcified granules
   (c) Contain hair cells
   (d) Contain endolymph which communicates with that in semicircular canals and cochlea
   (e) Can initiate reflex changes in muscle tone

309 The rods in the retina:
   (a) Contain visual pigment which is more sensitive to red than to blue light
   (b) Are rendered insensitive by ordinary daylight
   (c) Are more widely distributed over the retina than are cones
   (d) Reflect red light more than blue light
   (e) Comprise about 20% of foveal receptor cells

310 Cones:
   (a) Are found in the most superficial layer of the retina
   (b) Show a graded depolarization in response to light
   (c) Contain pigments which are more light sensitive than the rod pigment
   (d) Contain pigments which are most affected by yellow–green light
   (e) Are absent in an individual with colour blindness

311 The receptor cells serving taste:
   (a) Are confined to the tongue
   (b) Are stimulated when chemicals diffuse through the overlying epithelium to reach them
   (c) Are primary sensory neurones
   (d) Are histologically different for the four primary taste modalities
   (e) For sweetness are more common at the tip than at the back of the tongue

312 Sound waves:
   (a) Are quantified on the decibel scale which is logarithmic
   (b) With an intensity of 0 dB are inaudible
   (c) May have an intensity of –10 dB
   (d) With an intensity of 90 dB are usually painful
   (e) Are heard as a note one octave higher when their frequency increases eight-fold

313 The frequency of impulses generated by receptors in a utricle is:
   (a) Related to the orientation of the head
   (b) Higher during travel at 100 than at 20 miles/h
   (c) Reduced in the weightless conditions in outer space
   (d) Inversely related to the frequency being generated by the opposite utricle
   (e) Related to the frequency being generated by receptors in the semicircular canals

308 **(a)** **True**   They respond to linear acceleration.
    (b) **True**   These otoliths enable it to respond to linear acceleration.
    (c) **True**   These are stimulated by forces acting on the otoliths.
    (d) **True**   They are all parts of the membranous labyrinth, filled with endolymph.
    (e) **True**   Muscle tone is redistributed so that the body can withstand gravitational stresses.

309 **(a)** **False**   Rhodopsin absorbs blue–green light with a wavelength around 500 nm.
    (b) **True**   Nearly all rhodopsin is broken down (bleached) in daylight.
    (c) **True**   The field of vision using rods is greater than that using cones.
    (d) **True**   Rhodopsin is a red pigment.
    (e) **False**   There are no rods in the fovea

310 **(a)** **False**   They are in the layer furthest from the vitreous humour.
    (b) **False**   They hyperpolarize in light due to closure of Na^+ channels in the membrane.
    (c) **False**   Rhodopsin is the most sensitive of the pigments.
    (d) **True**   Cone pigments absorb light at wavelengths of 440, 535 and 565 nm; yellow–green light shows up relatively well in dim light.
    (e) **False**   Colour-blind individuals lack one of the three cone systems.

311 **(a)** **False**   They are found also in the soft palate, pharynx and larynx.
    (b) **False**   The microvilli on top of receptors protrude though taste pores into the buccal cavity.
    (c) **False**   They are receptor cells which synapse with primary sensory neurones.
    (d) **False**   They look alike.
    (e) **True**   Sweet sensation is experienced at the front of the tongue; bitterness at the back.

312 **(a)** **True**   Loudness is related to the logarithm of sound intensity.
    (b) **False**   0 dB is the average threshold for hearing.
    (c) **True**   These sounds may be heard by people whose hearing ability is above average.
    (d) **False**   The threshold for pain is around 120 dB.
    (e) **False**   They sound one octave higher when their frequency is doubled.

313 **(a)** **True**   This determines the plane of the gravitational pull on the otoliths.
    (b) **False**   The utricle is affected by acceleration, not velocity.
    (c) **True**   This can give rise to a form of travel sickness.
    (d) **False**   Often both respond in parallel.
    (e) **False**   They function independently; utricles respond to linear acceleration, semicircular canals to angular acceleration.

314 Aqueous humour:
  (a) Is produced by diffusion and active transport in the ciliary bodies
  (b) Pressure is close to mean arterial pressure
  (c) Formation depends on the enzyme carbonic anhydrase
  (d) Is absorbed into veins at the junction of the iris and the cornea
  (e) Is more easily absorbed when the pupil is widely dilated

315 Rhodopsin, the pigment of the rods, is:
  (a) A purple pigment
  (b) Highly absorbent of red light
  (c) Most sensitive to violet light
  (d) Regenerated in the dark
  (e) Least sensitive to red light

316 In the visual field of the left eye, an object:
  (a) In the upper temporal quadrant is detected in the lower nasal quadrant of the retina
  (b) At the centre of the field of vision is detected in the optic disc
  (c) Focused on the blind spot is in the nasal half of the visual field
  (d) In the temporal half generates impulses which travel in the left optic tract
  (e) In the nasal half is more likely to be perceived in binocular vision than one in the temporal half

317 The basilar membrane:
  (a) Is broader at the base of the cochlea than at the apex
  (b) Vibrations stimulate receptors to generate impulses at the frequencies of the applied sounds
  (c) At the base of the cochlea vibrates only to incoming high-frequency sounds
  (d) In the apical region vibrates only to incoming sounds of low frequency
  (e) Can be made to vibrate by pressure waves travelling through skull bone

318 Taste receptors:
  (a) For sour taste predominate at the sides of the tongue
  (b) May respond to more than one modality of stimulus
  (c) Give rise to a sour taste when stimulated by hydrogen ions
  (d) Cannot detect small (<10%) differences in the concentration of taste-evoking chemicals
  (e) Are more excited by warm solutions than by cold

314 (a) **True**   Its crystalloid composition is not identical to that of plasma.
  (b) **False**   Above 20 mmHg is abnormal and suggests glaucoma.
  (c) **True**   Inhibitors of this enzyme reduce formation and are used in treatment of glaucoma.
  (d) **True**   This area with its overlying trabeculae is referred to as the canal of Schlemm.
  (e) **False**   The iris then blocks access to the canal of Schlemm.

315 (a) **False**   It is red.
  (b) **False**   It is red because it reflects red light selectively.
  (c) **False**   It is most sensitive to blue–green light (around 500 nm).
  (d) **True**   In the dark, retinene and scotopsin combine to form rhodopsin.
  (e) **True**   This is because it reflects the red light.

316 (a) **True**   The image is inverted and reversed with respect to the object.
  (b) **False**   The point focused upon is detected at the macula (fovea).
  (c) **False**   The optic disc is medial to the fovea, so the blind spot is in the temporal part of the field of vision.
  (d) **False**   Impulses related to the temporal region of the left field of vision cross to the right at the optic chiasma.
  (e) **True**   The visual fields of the two eyes overlap, apart from the outer temporal areas.

317 (a) **False**   The reverse is true.
  (b) **False**   Nerves cannot transmit impulses at the top frequencies detectable by ear, about 20 000 Hz.
  (c) **False**   At the base, the basilar membrane vibrates to high- and low-frequency sound waves.
  (d) **True**   High-frequency sounds are damped out before they reach the apex.
  (e) **True**   This supplements the normal ossicular conduction, especially for loud sounds.

318 (a) **True**   Receptors for salt taste predominate on the anterior dorsum of the tongue.
  (b) **True**   Recording from single taste receptors demonstrates that a single receptor can respond to more than one modality.
  (c) **True**   All acids taste sour.
  (d) **True**   Taste receptors are poor at discriminating between intensities; a concentration difference of more than 30% is needed for discrimination.
  (e) **True**   Food flavour is accentuated when hot; unpleasant medicine is less offensive when cold.

319 An audiogram:
   (a) Is a plot of hearing loss (or hearing ability) against sound frequency
   (b) Showing equal impairment of air and bone conduction suggests conductive deafness
   (c) Showing hearing loss at low frequencies for air conduction suggests ear drum damage
   (d) Showing loss at 8000 Hz for air and bone conduction suggests basal cochlear damage
   (e) Showing hearing loss at the lower frequencies is typical of an elderly person

320 Poor balance is more likely when there is:
   (a) Semicircular canal rather than cochlear damage
   (b) Impairment of basilar rather than carotid artery blood flow
   (c) Spinothalamic tract rather than posterior column damage
   (d) Dim rather than bright light
   (e) Recent rather than long-standing destruction of one labyrinth

321 In someone with shortsightedness (myopia):
   (a) The eye tends to be longer than average from lens to retina
   (b) A convex lens is required to correct the refractive error
   (c) Close vision is affected more than distance vision
   (d) The near-point is farther than normal from the eye
   (e) A circular object tends to appear oval

322 Colour blindness:
   (a) Results from inability to detect one of the three primary light colours: red, yellow and blue
   (b) Where red and green are indistinguishable is due to failure of red and green cone systems
   (c) In which no colours can be detected is due to failure of all the cone systems
   (d) Is more common in women than men
   (e) Is a disability linked to the Y-chromosome

323 Local application of atropine to the eye causes:
   (a) Dilation of the pupil
   (b) The near-point for clear vision to move closer to the eye
   (c) Inability to focus on objects at infinity
   (d) Reduced drainage of aqueous humour
   (e) Difficulty in looking upwards

319 (a) **True** It is obtained using an audiometer.
   (b) **False** In conductive deafness air conduction, but not bone conduction, is impaired.
   (c) **True** This is an example of conductive deafness.
   (d) **True** This can be caused by acoustic trauma, e.g. in heavy industry.
   (e) **False** Hearing loss in the elderly (presbycusis) particularly affects higher frequencies.

320 (a) **True** The cochlea does not contribute sensory information needed for balance.
   (b) **True** The basilar artery supplies brain-stem areas particularly concerned with balance.
   (c) **False** The posterior columns transmit proprioceptive information needed for balance.
   (d) **True** Vision can compensate for loss of proprioception.
   (e) **True** Abrupt loss of input causes severe disturbance followed by gradual adaptation.

321 (a) **True** Hence distant objects are focused in front of the retina.
   (b) **False** A concave lens is required.
   (c) **False** It is distant objects that appear out of focus.
   (d) **False** Myopic people can focus on objects closer to the eye than normal people.
   (e) **False** This is caused by an asymmetrical cornea (astigmatism).

322 (a) **True** One or more of the three types of cone fails to function.
   (b) **False** It is due to failure of one of the two systems.
   (c) **False** It is due to the presence of only one functioning cone system.
   (d) **False** It is 20 times more common in men.
   (e) **False** Its linkage to the X-chromosome explains its greater frequency in men.

323 (a) **True** This paralyses the cholinergic constrictor fibres.
   (b) **False** The ciliary muscle needed for accommodation has cholinergic innervation.
   (c) **False** When focusing at infinity, the ciliary muscle is at rest.
   (d) **True** The iris tends to obstruct the canal of Schlemm in the corneoscleral angle.
   (e) **False** Extraocular muscles are not affected by atropine.

324 In the middle ear:
  (a) Destruction of the auditory ossicles abolishes hearing
  (b) Paralysis of the auditory muscles makes sounds more faint
  (c) Immobilization of the stapes causes greater deafness than removal of the ossicles
  (d) Air pressure is normally atmospheric pressure
  (e) The round window moves reciprocally with the oval window

325 Interruption of the visual pathway in the:
  (a) Left optic tract causes blindness in the right visual field (right homonymous hemianopia)
  (b) Optic chiasma causes blindness in the nasal half of each visual field (binasal hemianopia)
  (c) Left optic radiation causes loss of vision to the right
  (d) Occipital cortex causes loss of the light reflex
  (e) Occipital cortex causes loss of central vision with preservation of peripheral vision

326 Squinting (strabismus) may result from:
  (a) Poor vision in one eye in childhood
  (b) A refractive error in childhood
  (c) Central suppression of vision in one eye in childhood
  (d) Damage to the cerebellum
  (e) Damage to the internal capsule

327 Impairment of the sense of smell:
  (a) May be confined to certain odours only
  (b) May occur in hydrocephalus
  (c) Is likely after thalamic damage
  (d) Can be caused by inflammation of the nasal mucosa
  (e) Is a recognized effect of temporal lobe tumour

328 Involuntary oscillatory eye movements (nystagmus):
  (a) Do not occur in normal people
  (b) May result from cochlear disease
  (c) Occur in cerebellar disease
  (d) Occur when cold fluid is run into one external ear canal
  (e) Do not affect acuity of vision

329 Typical effects of ageing on the special senses include gradual loss of:
  (a) Near vision
  (b) Olfactory sensitivity (hyposmia)
  (c) 90% of the accommodative power of the lens during the life-span
  (d) Hearing affecting bone and air conduction similarly
  (e) Hearing affecting high and low frequencies similarly

324 (a) False Sound can still be transmitted by bone conduction.
   (b) False These muscles damp vibration of the ossicles reflexly in response to loud noises.
   (c) True Immobilization prevents the oval window vibrating and causes severe deafness.
   (d) True Air pressure is atmospheric, due to the patency of the pharyngotympanic tubes.
   (e) True Fluid (endolymph) cannot be compressed; as the oval window moves in, the round window moves out.

325 (a) True The left half of each retina is concerned with vision to the right and impulses from them travel in the left optic tract.
   (b) False The crossing fibres come from the nasal half of each retina and are responsible for temporal vision; bitemporal hemianopia results.
   (c) True As with damage to the left optic tract.
   (d) False This is a brain-stem reflex.
   (e) False The reverse is true because the fovea is bilaterally represented in the cortex.

326 (a) True This impairs fixation of the eye concerned.
   (b) True This impairs vision in the eye concerned to give poor fixation.
   (c) False This is a consequence not a cause of squint.
   (d) False This may cause involuntary oscillatory movements (nystagmus) but not squinting.
   (e) True This may cause a paralytic squint due to damage to the oculomotor tracts.

327 (a) True If only some of the many receptor types involved in olfaction are lost.
   (b) True Due to damage to the olfactory nerves by distortion of the cranium.
   (c) False Smell pathways do not pass through the thalamus.
   (d) True This can prevent odours reaching the receptor cells.
   (e) False It may indicate a frontal lobe tumour.

328 (a) False They occur when a normal person stops rotating.
   (b) False Disease of the semicircular canals may cause nystagmus.
   (c) True Nystagmus is an ataxia of eye fixation.
   (d) True Due to cooling of fluid in the adjacent semicircular canal.
   (e) False The rapid eye movements tend to make vision blurred.

329 (a) True Recession of the near-point is typical of the ageing eye (presbyopia); vision at 20–30 cm deteriorates.
   (b) True It affects over 70% of elderly people.
   (c) True It falls from 10–15 dioptres in childhood to 5–10 at 30 and to about 1 dioptre at 70.
   (d) True It is a sensorineural deafness (presbycusis).
   (e) False High-pitched sounds are more affected.

330 A child who focuses an object on the fovea of the left eye and on the temporal side of the fovea in the right eye is likely to have:
  (a) A divergent squint
  (b) A refractive error
  (c) Suppression of vision in the left rather than in the right eye
  (d) No suppression of vision in one eye if the left eye is covered for part of each day
  (e) Greater suppression of vision in one eye if exposed to exercises requiring binocular vision

331 In unilateral vestibular disease, typical features include:
  (a) The sensation that the external world is revolving
  (b) Prolonged nystagmus when cold water is placed in the external auditory meatus on the affected side
  (c) A tendency to stagger when walking
  (d) A tendency to fall in the dark
  (e) Nausea and vomiting

332 Impairment of visual acuity in bright light can be explained by:
  (a) Random light scattering when there is deficient pigmentation of the eye due to albinism
  (b) Random light scattering when there is asymmetrical corneal curvature due to astigmatism
  (c) Random light scattering in the cornea when there is vitamin A deficiency
  (d) Impairment of rod function when there is vitamin A deficiency
  (e) Inability to alter the focal length of the lens when a cataract is present

333 In longsightedness (hypermetropia):
  (a) Objects at infinity cannot be focused sharply on the retina
  (b) Objects at the usual near-point are focused behind the retina
  (c) Ciliary muscle contracts more strongly to bring objects in mid-visual range into clear focus
  (d) The range of unblurred vision (near-point to far-point) is greater than normal
  (e) The near-point can be brought closer to the eye by the use of a biconcave lens

330 (a) **False**   The right eye is converging, making it a convergent squint.
     (b) **True**   Refractive errors are a common cause of squinting.
     (c) **False**   Suppression of vision tends to occur in the non-fixing eye.
     (d) **True**   Covering the 'good' eye helps to preserve vision in the non-fixing eye.
     (e) **False**   Such exercises help in the development of binocular vision.

331 (a) **True**   Unbalanced vestibular input causes this sensation (vertigo).
     (b) **False**   In this 'caloric' test, reduction in nystagmus duration indicates vestibular abnormality.
     (c) **True**   Due to inappropriate information affecting brain areas controlling balance.
     (d) **True**   Compensating visual stimuli are then eliminated.
     (e) **True**   Unbalanced, excessive or reduced vestibular inputs cause nausea and vomiting as seen in sea-sickness and space travel sickness.

332 (a) **True**   Normally absorption of light by dark pigment in the choroid prevents back-scattering of light into the retina.
     (b) **False**   There is a refractive error but not random light scattering.
     (c) **True**   Lack of vitamin A leads to keratin deposition in corneal epithelium (xerophthalmia).
     (d) **False**   Rod function does not determine acuity in bright light.
     (e) **False**   Impairment of acuity with cataract is due to random scattering by lens opacities.

333 (a) **False**   This is true of shortsightedness (myopia).
     (b) **True**   The eye is usually shorter than normal.
     (c) **True**   This distance is closer than usual to the hypermetrope's near-point.
     (d) **False**   It is less than normal; the far-point stays at infinity but the near-point is further from the eye.
     (e) **False**   A convex lens is required to augment the power of the eye's refracting system.

334 Hydrostatic pressure in renal glomerular capillaries:
- (a) Is lower than pressure in efferent arterioles
- (b) Rises when afferent arterioles constrict
- (c) Is higher than in most other capillaries at heart level
- (d) Falls by 10% when arterial pressure falls by 10%
- (e) Falls as the oncotic pressure rises along the length of the capillary

335 Tubular reabsorption of a filtered substance is likely to be active rather than passive if its:
- (a) Concentration in the tubular fluid is lower than in peritubular capillary blood
- (b) Excretion is increased by cooling the kidney
- (c) Renal clearance is lower than that of inulin
- (d) Renal clearance rises at high plasma levels
- (e) Urinary excretion rate:plasma concentration ratio is the same as for glucose

336 The renal clearance of a substance:
- (a) Is inversely related to its urinary concentration, U
- (b) Is directly related to the rate of urine formation, V
- (c) Is directly related to its plasma concentration, P
- (d) Is expressed in units of volume per unit time
- (e) Must fall in the presence of metabolic poisons

337 In fluid in the distal part of the proximal convoluted tubule:
- (a) Urea concentration is higher than in Bowman's capsule
- (b) pH is less than 6 when the kidneys are excreting an acid urine
- (c) Glucose concentration is similar to that in plasma
- (d) Osmolality is about 25% that of glomerular filtrate
- (e) Bicarbonate concentration is lower than in plasma

338 Renal tubules normally reabsorb:
- (a) More water every hour than the entire plasma volume
- (b) All filtered HCO_3^- in respiratory acidosis
- (c) All filtered amino acids
- (d) All filtered plasma proteins
- (e) More K^+ than Cl^-

339 As plasma glucose concentration rises above normal, glucose:
- (a) Filtration increases linearly
- (b) Transport maximum increases linearly
- (c) Clearance increases linearly
- (d) Reabsorption increases and then levels off
- (e) Excretion increases and then decreases

334 (a) **False** It must be higher to maintain blood flow.
   (b) **False** The pressure drop across the afferent arterioles increases as they constrict.
   (c) **True** The afferent arterioles offer relatively little resistance.
   (d) **False** Redistribution of renal vascular resistance due to autoregulation tends to maintain glomerular hydrostatic pressure and hence filtration.
   (e) **True** Hydrostatic pressure falls due to vascular resistance; oncotic pressure rises due to loss of protein-poor filtrate.

335 (a) **True** This suggests transportation into the blood against a concentration gradient.
   (b) **True** Cooling impairs active metabolic processes.
   (c) **False** This indicates reabsorption but not whether it is active (e.g. glucose) or passive (e.g. urea).
   (d) **True** This suggests saturation of a carrier system.
   (e) **True** Anything filtered in glomeruli having zero clearance must be actively reabsorbed.

336 (a) **False** It is directly related to urinary concentration.
   (b) **True** Clearance tends to fall at low urinary flow rates.
   (c) **False** It is inversely related to plasma concentration.
   (d) **True** Clearance = UV/P in units of volume/unit time.
   (e) **False** It rises if the substance is normally reabsorbed by an active process.

337 (a) **True** Due to reabsorption of water.
   (b) **False** Acidification occurs mainly in the distal convoluted tubule.
   (c) **False** Most or all of the glucose is reabsorbed before the end of the proximal tubule.
   (d) **False** Osmolality changes little in the proximal convoluted tubule
   (e) **True** Like glucose, HCO_3^- is usually completely reabsorbed in the proximal tubule.

338 (a) **True** About 99% of the glomerular filtrate.
   (b) **True** This plus HCO_3^- manufactured in the kidney compensates the respiratory acidosis.
   (c) **True** These are filtered but do not appear in normal urine.
   (d) **True** Again some are filtered but do not appear in urine.
   (e) **False** About 20 times as much chloride as potassium is filtered.

339 (a) **True** Filtration rate is directly proportional to concentration.
   (b) **False** Transport maxima are constants.
   (c) **False** It remains at zero until the T_m is reached and then it rises linearly.
   (d) **True** It levels off after T_m glucose is reached.
   (e) **False** It is initially zero and then rises linearly.

340 A substance is being secreted by the renal tubules if its:
  (a) Clearance rate is greater than 150 ml/min
  (b) Concentration is higher in arterial than in renal venous blood
  (c) Excretion rate is increased by tubular enzyme poisons
  (d) Concentration rises along the proximal convoluted tubule
  (e) Concentration in urine is greater than in plasma

341 In the nephron, the osmolality of fluid in the:
  (a) Tip of the loop of Henle is less than that of plasma
  (b) Bowman's capsules is less than that in the distal tubules
  (c) Collecting duct rises when vasopressin is being secreted
  (d) Proximal convoluted tubule rises along its length
  (e) Medullary interstitium can exceed 1 osmol/litre

342 Transport maximum (T_m)-limited reabsorption of a substance implies that its:
  (a) Reabsorption is active
  (b) Reabsorption is critically related to tubular transit time
  (c) Reabsorption is complete below a certain threshold load
  (d) Renal clearance falls with its plasma concentration
  (e) Excretion rate is zero until its T_m value is reached

343 When a patient's mean arterial blood pressure falls by 50%:
  (a) Renal blood flow falls by less than 10%
  (b) Glomerular filtration falls by about 50%
  (c) There is an increase in the circulating aldosterone level
  (d) Renal vasoconstriction occurs
  (e) Urinary output ceases

344 The cells of the distal convoluted tubule:
  (a) Reabsorb about 50% of the water filtered by the glomeruli
  (b) Secrete hydrogen ions into the tubular lumen
  (c) Form NH_4^+ ions
  (d) Reabsorb sodium in exchange for hydrogen or potassium ions
  (e) Determine the final composition of urine

345 If, during an infusion of para-aminohippuric acid, peripheral venous plasma
  PAH level is 0.02 mg/ml (not above renal threshold), urinary PAH level is 16
  mg/ml and urinary flow rate 1 ml/min, then the:
  (a) PAH level in renal venous blood must exceed 0.02 mg/ml
  (b) PAH level in renal arterial blood must be about 0.02 mg/ml
  (c) PAH level in glomerular filtrate must be about 0.02 mg/ml
  (d) Renal plasma flow is nearer 800 than 1000 ml/min
  (e) Renal blood flow is nearer 1300 than 1500 ml/min if the haematocrit is
     0.40

340 **(a)** **True** A clearance value above the glomerular filtration rate (about 140 ml/min) indicates secretion.
(b) **True** Some of the unfiltered fraction must have been secreted.
(c) **False** This suggests that the substance is normally reabsorbed by an active process.
(d) **False** This can be explained by water reabsorption.
(e) **False** Again, this can be explained by a greater reabsorption of water.

341 **(a)** **False** This fluid is hypertonic because of countercurrent concentration.
(b) **False** Distal tubular fluid is hypotonic.
(c) **True** Vasopressin (ADH) promotes water, but not salt, reabsorption in collecting ducts.
(d) **False** The fluid remains isotonic with plasma.
(e) **True** It can be about four times that of plasma.

342 **(a)** **True** T_m-limited reabsorption is one type of active tubular reabsorption.
(b) **False** This applies to the other type, gradient-time-limited reabsorption.
(c) **True** As with glucose.
(d) **False** Clearance is zero at all levels below the threshold.
(e) **True** The concept applies also to amino acids and proteins.

343 **(a)** **False** Autoregulation cannot compensate for such large falls.
(b) **False** It falls to about zero when glomerular capillary pressure falls below the sum of intracapsular pressure plus plasma oncotic pressure – about 40 mmHg.
(c) **True** Due to release of renin and angiotensin formation, aldosterone is secreted.
(d) **True** Reflex sympathetic vasoconstriction due to decreased baroreceptor stimulation.
(e) **True** When glomerular filtration stops, urinary output stops.

344 **(a)** **False** About 80% of the filtered water is reabsorbed before it reaches the distal tubules.
(b) **True** The rate is related to acid–base requirements.
(c) **True** By conversion of glutamine to glutamate; NH_3 is a buffer for the H^+ being excreted.
(d) **True** H^+ secretion is related to the body's acid–base balance.
(e) **False** Further modification takes place in the collecting ducts.

345 **(a)** **False** The renal venous blood level would be negligible.
(b) **True** Since PAH is excreted only by the kidneys, the PAH level in peripheral venous blood determines the level entering the renal arteries.
(c) **True** Since PAH is freely filtered.
(d) **True** Flow = PAH clearance = UV/P = 16 × 1/0.02 = 800 ml/min.
(e) **True** Blood flow = plasma flow/0.6 = 1333 ml/min.

346 Renal blood flow falls:
   (a) About 10% when arterial pressure falls 10% below normal
   (b) About 5% when metabolic activity in the kidney falls by 5%
   (c) During emotional stress
   (d) After moderate haemorrhage
   (e) Gradually from the inner medulla to the outer cortex per unit weight of tissue

347 Urea:
   (a) And glucose have similar molar concentrations in normal blood
   (b) Concentration rises in tubular fluid as the glomerular filtrate passes down the nephron
   (c) Is actively secreted by the renal tubular cells into the tubular fluid
   (d) Concentration in blood may rise ten-fold after a high protein meal
   (e) Causes a diuresis when its blood concentration is increased

348 Voluntary micturition:
   (a) Depends on the integrity of a lumbar spinal reflex arc
   (b) Is not possible after sensory denervation of the bladder
   (c) Involves stimulation of the detrusor muscle in the bladder by autonomic sympathetic nerves
   (d) Is normally accompanied by some reflux of bladder contents into the ureters
   (e) Is inhibited during ejaculation

349 The proximal convoluted tubules:
   (a) Reabsorb most of the sodium ions in glomerular filtrate
   (b) Reabsorb most of the chloride ions in glomerular filtrate
   (c) Reabsorb most of the potassium ions in glomerular filtrate
   (d) Contain juxtaglomerular cells which secrete renin
   (e) Contain the main target cells for antidiuretic hormone

350 The renal clearance of:
   (a) Inulin provides an estimate of glomerular filtration rate
   (b) Chloride increases after an injection of aldosterone
   (c) PAH falls when the PAH load exceeds the T_m for PAH
   (d) Urea is lower than that of inulin
   (e) Inulin is independent of its plasma concentration

346 (a) False Due to autoregulation, flow changes little with small changes in perfusion pressure.

(b) False Normal renal blood flow is vastly in excess of its metabolic requirements.

(c) True Due to sympathetic vasoconstrictor nerves and circulating catecholamines.

(d) True A reflex response to the fall in blood pressure so caused.

(e) False Cortical flow is 10–20 times higher than medullary flow.

347 (a) True Both are around 5 mmol/litre.

(b) True The urinary concentration of urea is many times that in plasma.

(c) False 50% of the filtered urea is passively reabsorbed; the rise in tubular concentration can be explained by the reabsorption of water.

(d) False It rises but would not double in concentration.

(e) True It causes an osmotic diuresis.

348 (a) False The reflex centres are in the sacral cord; their activity is modulated by higher centres.

(b) True This breaks the reflex arc.

(c) False Parasympathetic nerves are motor to the detrusor muscle.

(d) False Valves where the ureters enter the bladder do not allow such reflux.

(e) True During ejaculation, sympathetic activity constricts the bladder neck sphincter and prevents retrograde ejaculation of semen into the bladder.

349 (a) True About 80% of the filtered sodium is actively absorbed in the proximal tubules.

(b) True Negatively charged chloride ions follow the sodium.

(c) True Most of the potassium is reabsorbed in the proximal tubule; some is re-excreted in the distal tubules in exchange for sodium.

(d) False These are found where the distal tubule makes contact with the afferent arteriole.

(e) False This hormone acts mainly on distal parts of the nephron.

350 (a) True Inulin is freely filtered but not reabsorbed or secreted in the tubules; Therefore the amount excreted in the urine equals the amount filtered at the glomerulus.

(b) False Aldosterone increases Na^+ and Cl^- reabsorption and so reduces their clearance.

(c) True At high plasma levels, the T_m for PAH is exceeded and PAH is not completely cleared in one passage through the kidney.

(d) True About 60 compared with 120 ml/min; half the filtered urea is passively reabsorbed.

(e) True The amount filtered is the amount excreted.

351 The collecting ducts in the kidney:
  (a) Can actively transport water molecules into the urine
  (b) Are the site of most of renal water reabsorption
  (c) Are rendered impermeable to water by antidiuretic hormone (ADH)
  (d) Pass through a region of exceptional hypertonicity
  (e) Determine to a large extent the final osmolality of urine

352 Aldosterone:
  (a) Is a steroid hormone secreted by the adrenal medulla
  (b) Production ceases following removal of the kidneys and their juxtaglomerular cells
  (c) Production decreases in treatment with drugs which block angiotensin converting enzyme
  (d) Secretion results in increased potassium reabsorption by the nephron
  (e) Secretion results in a fall in urinary pH

353 As fluid passes down the proximal convoluted tubule, there is a fall of more than 50% in the:
  (a) Concentration of sulphate ions
  (b) Concentration of sodium ions
  (c) Concentrations of amino acids
  (d) Concentration of potassium ions
  (e) Rate of filtrate flow in the tubules

354 In normal healthy people, urinary:
  (a) Specific gravity ranges from 1.010 to 1.020
  (b) Osmolality ranges from 200 to 400 mosmol/litre
  (c) Colour is due to small quantities of bile pigments
  (d) pH falls as dietary protein rises
  (e) Calcium excretion is increased by parathormone

355 Aldosterone secretion tends to raise the volume of:
  (a) Plasma
  (b) Interstitial fluid
  (c) Intracellular fluid
  (d) Urine
  (e) Cerebrospinal fluid

351 (a) False Active transport of water has not been described in the body.

(b) False 80% of the water in glomerular filtrate is reabsorbed in the proximal tubules.

(c) False Conversely, they are rendered permeable to water by ADH.

(d) True Osmolality in the inner medullary interstitium can exceed 1 osmol/litre.

(e) True By determining the amount of water reabsorbed as the glomerular filtrate passes through the hypertonic medullary interstitium.

352 (a) False It is a steroid as its name suggests but is secreted by the adrenal cortex.

(b) False Aldosterone is secreted in response to ACTH secretion, high K^+ intake, heart failure, etc., in addition to activity in the renin/angiotensin system.

(c) True ACE inhibitor drugs tend to reduce the level of angiotensin II which stimulates the adrenal cortex to produce angiotensin.

(d) False It increases potassium secretion in exchange for sodium.

(e) True It increases H^+ secretion, also in exchange for sodium.

353 (a) False Sulphate concentration rises since more water than sulphate is reabsorbed.

(b) False It is little changed, since similar proportions of sodium and water are reabsorbed.

(c) True These are completely reabsorbed by active transport.

(d) False Potassium is reabsorbed in proportion to water.

(e) True Due to reabsorption of about 80% of the water.

354 (a) False It may range from 1.004 to 1.040.

(b) False It may range from 100 to 1000 mosmol/litre.

(c) False It is due to 'urochrome', a pigment of uncertain origin.

(d) True Dietary proteins lead to acid residues such as sulphates and phosphates.

(e) True More calcium is filtered due to the raised blood level, so more is excreted.

355 (a) True By retention of sodium chloride and water in the extracellular fluid compartment.

(b) True This, like plasma, is a subcompartment of the extracellular fluid.

(c) False The sodium chloride/water retention is confined to the extracellular compartment.

(d) False It reduces it by retaining salt and water.

(e) False CSF is a secretion classified as transcellular fluid; it is not a subcompartment of ECF.

356 The renal clearance of:
  (a) Bicarbonate is similar to that of glucose
  (b) PAH is nearer 600 than 1200 ml/min in the average adult
  (c) Creatinine provides an estimate of renal plasma flow
  (d) Phosphate is decreased by parathormone
  (e) Protein is normally zero

357 Potassium:
  (a) Is actively secreted in the distal convoluted tubule
  (b) Is reabsorbed in the proximal convoluted tubule
  (c) Deficiency favours hydrogen ion secretion in the distal tubule
  (d) Excretion is determined largely by potassium intake
  (e) Blood levels tend to rise in patients with acute renal failure taking a normal diet

358 Secretion of renin:
  (a) Occurs in the stomach mucosa during infancy
  (b) Is stimulated by the hormone angiotensin I
  (c) Leads to raised levels of angiotensin II in the blood
  (d) Is stimulated by a fall in extracellular fluid volume
  (e) Inhibits ACTH secretion by the pituitary gland

359 In chronic renal failure:
  (a) Glomerular filtration rate may fall by 70% before the condition gives rise to symptoms
  (b) The specific gravity of the urine tends to be elevated, e.g. about 1.030
  (c) Blood P_{CO_2} tends to be low
  (d) Ionized calcium levels in the blood tend to be high
  (e) Anaemia is common

360 Diabetes insipidus (deficiency of antidiuretic hormone) causes a fall in the:
  (a) Osmolality of the urine
  (b) Reabsorption of water from the proximal tubules
  (c) Extracellular but not intracellular fluid volume
  (d) Extracellular fluid osmolality
  (e) Intracellular fluid osmolality

356 (a) True   Both are usually totally reabsorbed so their renal clearance is about zero.

   (b) True   PAH clearance is a measure of renal plasma flow, not renal blood flow.

   (c) False   It provides an estimate of the glomerular filtration rate since the amount filtered is close to the amount excreted.

   (d) False   Phosphate clearance is increased by parathormone and lowers the blood phosphate level.

   (e) True   Small amounts of protein are filtered but reabsorbed.

357 (a) True   In exchange for sodium ions.

   (b) True   It is reabsorbed passively down the gradient created by Na^+ and H_2O reabsorption.

   (c) True   Potassium and hydrogen compete for secretion in exchange for sodium.

   (d) True   Thus potassium balance is maintained.

   (e) True   In acute renal failure, the failure to excrete the potassium intake leads to high blood levels which can compromise the performance of the heart.

358 (a) False   *Rennin* is the enzyme secreted by infants' gastric mucosa which curdles milk.

   (b) False   Renin promotes angiotensin I formation from a circulating precursor.

   (c) True   Angiotensin I is converted to angiotensin II by a converting enzyme in the lungs.

   (d) True   Renin's action helps to restore this volume.

   (e) False   There is no direct feedback between the two systems.

359 (a) True   The kidneys have a large functional reserve.

   (b) False   Renal ability to concentrate urine is impaired; the range of specific gravity decreases, converging towards that of protein-free plasma, 1.010.

   (c) True   Poor excretion of acid residues causes metabolic acidosis which stimulates ventilation.

   (d) False   Ionized calcium levels fall due to retention of phosphate ions and failure of renal activation of vitamin D.

   (e) True   Most often it is due to deficiency of erythropoietin.

360 (a) True   Due to failure of the kidneys to reabsorb sufficient water.

   (b) False   Reabsorption of water in proximal tubules is normal since it depends on the active reabsorption of sodium; reabsorption in the collecting ducts is affected.

   (c) False   Both fluid compartments are depleted in volume.

   (d) False   It rises due to depletion of water but not salt.

   (e) False   Both compartments show the same raised osmolality; osmotic gradients are effective in moving water at cell membranes.

361 The cystometrogram shows:
- (a) A plot of bladder pressure on the ordinate axis against bladder volume on the abscissa
- (b) Little rise in pressure with rise in volume at low bladder volumes
- (c) A steep rise in pressure when volume rises above 100 ml
- (d) That females generate higher pressures during micturition than males
- (e) That patients with chronic urinary tract obstruction can generate higher than normal micturition pressures

362 Treatment with an aldosterone antagonist causes a fall in:
- (a) Urine volume
- (b) Body potassium
- (c) Body sodium
- (d) Blood volume
- (e) Blood viscosity

363 Dialysis fluid used in the treatment of renal failure should contain the normal plasma levels of:
- (a) Urea
- (b) Potassium
- (c) Osmolality
- (d) Plasma proteins
- (e) Hydrogen ions

364 Long-standing obstruction of the urethra may cause:
- (a) Enlargement of the prostate gland
- (b) Hypertrophy of the bladder muscle
- (c) Dilation of the ureters
- (d) Reduction of the glomerular filtration rate
- (e) An increase in residual volume in the bladder

365 Emptying of the bladder may be less effective if:
- (a) The sympathetic nerves carrying afferent information from bladder to spinal cord are cut
- (b) The pelvic nerves are cut
- (c) Anticholinergic drugs are administered
- (d) Alpha adrenergic receptor agonists are administered
- (e) Beta adrenergic receptor agonists are administered

366 Renal transplantation for chronic renal failure in adults should:
- (a) Be covered by immunosuppression even when the donor is the recipient's identical twin
- (b) Raise postoperative glomerular filtration rate to the 10–20 ml/min level
- (c) Correct abnormal calcium metabolism
- (d) Correct anaemia
- (e) Abolish the need for further renal dialysis

361 (a) True Bladder pressure is measured while known volumes of fluid are run into it.

(b) True An example of receptive relaxation like that seen in the stomach.

(c) False The deflection usually occurs when around 500 ml is introduced.

(d) False The male urinary tract offers a higher 'peripheral resistance'.

(e) True The increased work load causes muscular hypertrophy which allows generation of higher micturition pressures.

362 (a) False It increases due to increased salt and water loss.

(b) False Body potassium rises since aldosterone normally increases its excretion.

(c) True Due to decreased sodium reabsorption.

(d) True Due to decreased extracellular fluid volume.

(e) False The viscosity increases as the haematocrit increases.

363 (a) False It should be urea free to provide a high concentration gradient for urea loss.

(b) False It should be lower to favour loss of potassium, which is elevated in renal failure.

(c) False It should be higher to reduce extracellular fluid volume and hence blood pressure.

(d) False Fluid transfer is governed by hydrostatic pressure and crystalloid osmolality gradients, not by colloid osmotic pressure gradients.

(e) True It should be buffered to prevent large pH changes.

364 (a) False This is a cause, not a consequence.

(b) True Due to the increased work it has to do.

(c) True Long-standing obstruction leads to urinary reflux when the uretero-vesical valves become incompetent.

(d) True Back-pressure in the ureters is transmitted to the nephrons and raises capsular pressure in the glomerulus.

(e) True This encourages urinary tract infection.

365 (a) False Sympathetic trunks carry pain afferents, not stretch receptor afferents to the cord.

(b) True These carry the stretch receptor afferents from the bladder and parasympathetic motor fibres to the bladder; the micturition reflex is lost.

(c) True These block the parasympathetic motor fibres to the detrusor muscle.

(d) True They contract the bladder sphincter muscle.

(e) True They tend to relax the detrusor muscle.

366 (a) False Donor and recipient have identical genes and immunological characteristics.

(b) False It should raise it to near normal, 120–150 ml/min.

(c) True This reverses the tendency towards demineralization of bone.

(d) True The transplanted kidney should supply the missing erythropoietin.

(e) True A healthy transplanted kidney should return all aspects of renal function to normal.

367 Drugs which interfere with active transport of sodium in the proximal tubule tend to increase:
   (a) Urine production
   (b) Plasma osmolality
   (c) Chloride excretion
   (d) Interstitial fluid volume
   (e) Plasma specific gravity

368 A drug which inhibits carbonic anhydrase decreases:
   (a) Bicarbonate formation and reabsorption in the kidney
   (b) Plasma bicarbonate levels
   (c) Blood pH
   (d) Urinary loss of potassium ions
   (e) Urinary volume and pH

369 A patient with chronic renal failure usually has an increased:
   (a) Blood urea
   (b) Blood uric acid
   (c) Creatinine clearance
   (d) Acid–base disturbance when he or she vomits
   (e) Acid–base problem on a high protein diet

370 Cutting the sympathetic nerves to the bladder may cause:
   (a) Difficulty in emptying the bladder
   (b) Loss of tone in the internal sphincter of the bladder
   (c) Loss of tone in the external sphincter of the bladder
   (d) Loss of pain sensation in the bladder
   (e) Infertility in the male

371 Sudden (acute) renal failure differs from gradual (chronic) renal failure in that:
   (a) Potassium retention tends to be more severe
   (b) Blood urea levels tend to be higher
   (c) Depression of bone marrow activity is less likely to occur
   (d) Metabolic acidosis is usually not a problem
   (e) Dietary protein restriction is unnecessary

372 In the treatment of someone with progressive renal failure:
   (a) Protein should be excluded from the diet
   (b) Water intake should be restricted to about 0.5 litre/day
   (c) The diet should be potassium free
   (d) Adequate dietary iron intake prevents anaemia
   (e) The calorific value of the diet should be gradually reduced

367 (a) **True**   By increasing salt and hence water loss.
 (b) **False**   This is regulated by ADH and the collecting ducts.
 (c) **True**   It passively follows the sodium being excreted.
 (d) **False**   This falls with the loss of salt and water.
 (e) **True**   Due to concentration of the proteins by removal of water.

368 (a) **True**   Carbonic anhydrase in tubular cells catalyses the combination of CO_2 and H_2O to form H_2CO_3 which ionizes into H^+ and HCO_3^+ ions.
 (b) **True**   This is determined mainly by renal bicarbonate formation.
 (c) **True**   This falls as the plasma bicarbonate level falls.
 (d) **False**   More K^+ is secreted by the tubules in exchange for sodium since there are fewer H^+ ions to compete with K^+ on the sodium/potassium exchange pump.
 (e) **False**   Failure to reabsorb HCO_3 results in an osmotic diuresis of alkaline urine.

369 (a) **True**   This is the standard way of making the diagnosis.
 (b) **True**   As with other end-products of protein digestion.
 (c) **False**   Creatinine clearance, a measure of GFR, is reduced in proportion to the severity of the renal failure.
 (d) **False**   Loss of the acid vomitus would improve the typical acidosis.
 (e) **True**   Proteins are a major source of the acid residues and toxic substances which accumulate in renal failure.

370 (a) **False**   It may cause increased frequency of micturition.
 (b) **True**   Sympathetic activity tends to raise sphincter tone.
 (c) **False**   This sphincter is supplied by somatic nerves.
 (d) **True**   Afferent pain fibres run with the sympathetic nerves.
 (e) **True**   Sympathetic fibres are necessary for closure of the internal sphincter of the bladder during ejaculation to prevent reflux of seminal fluid.

371 (a) **True**   Potassium retention is one of the greatest hazards of acute renal failure and may cause death from myocardial depression.
 (b) **False**   The blood urea level is determined by the severity of the condition, not by its rate of progression.
 (c) **False**   Both may depress the marrow and lower RBC, polymorph and platelet counts.
 (d) **False**   Both impair renal bicarbonate production.
 (e) **False**   Protein restriction is advisable in both cases.

372 (a) **False**   A low protein diet is helpful but some protein is needed to provide essential amino acids for tissue maintenance.
 (b) **False**   Water intake may need to be increased to about 3 litres to clear waste products since the kidneys cannot concentrate urine.
 (c) **False**   Potassium intake is required to replace potassium lost in urine.
 (d) **False**   Anaemia is due to bone marrow depression, not iron deficiency.
 (e) **False**   Sufficient dietary intake is needed to prevent excessive tissue protein catabolism.

373 A long-standing increase in arterial P_{CO_2} (respiratory acidosis) leads to an increase in:

(a) Renal bicarbonate formation
(b) Urinary ammonium salts
(c) Plasma potassium concentration
(d) The ratio of monohydrogen to dihydrogen phosphate in urine
(e) Urinary bicarbonate excretion

373 (a) True   Plasma HCO_3 rises to compensate for the raised P_{CO_2} in respiratory acidosis.

    (b) True   In acidosis, tubular cells excrete more NH_4^+ to buffer the additional H^+ ions being secreted.

    (c) True   The increased secretion of H^+ ions in exchange for Na^+ results in decreased secretion of K^+ ions.

    (d) False   The ratio decreases as hydrogen ions are taken up by the phosphate buffer system.

    (e) False   The urine remains bicarbonate free.

8 ENDOCRINE SYSTEM

374 In plasma, the half-life of:
   (a) A hormone is half the time taken for it to disappear from the blood
   (b) Insulin is between 5 and 10 h
   (c) Thyroxine is longer than that of insulin
   (d) Thyroxine is longer than that of triiodothyronine
   (e) Noradrenaline is longer than that of acetylcholine

375 During sleep there is a fall in the circulating level of:
   (a) Cortisol
   (b) Insulin
   (c) Adrenaline
   (d) Antidiuretic hormone
   (e) Growth hormone

376 Adrenocorticotrophic hormone (ACTH) secretion increases:
   (a) When the median eminence of the hypothalamus is stimulated
   (b) When aldosterone blood level falls
   (c) When cortisol blood levels fall
   (d) In bursts during the night as the normal hour of wakening approaches
   (e) Following severe trauma

377 Melatonin:
   (a) Is produced mainly in the intermediate lobe of the pituitary gland
   (b) Is synthesized in the body from serotonin (5-hydroxytryptamine)
   (c) Affects the level of pigmentation in human skin
   (d) Blood levels are highest during the night
   (e) Influences the secretion rates of pituitary hormones

378 Thyroid hormones, when secreted in excess, may cause an increase in the:
   (a) Peripheral resistance
   (b) Frequency of defaecation
   (c) Energy expenditure required for a given work load
   (d) Duration of tendon reflexes
   (e) Heart rate when cardiac adrenergic and cholinergic receptors are blocked

379 Aldosterone secretion is increased by an increase in plasma:
   (a) Volume
   (b) Osmolality
   (c) Potassium concentration
   (d) Renin concentration
   (e) ACTH concentration

374 **(a)** **False** It is the time it takes for the initial concentration to fall by half.

    (b) **False** It is much shorter (about 5 min); this allows more precise and continuous regulation of the blood glucose level.

    (c) **True** It is much longer since moment-to-moment regulation of its level is less critical.

    (d) **True** It is more highly protein bound, which appears to prolong its life.

    (e) **True** Acetylcholine is broken down almost immediately by cholinesterase.

375 **(a)** **True** The waking catabolic state changes to an anabolic state.

    (b) **True** Insulin secretion occurs mainly in association with meals.

    (c) **True** Adrenaline secretion is associated with stress.

    (d) **False** This rises as plasma osmolality rises; water is lost but not replaced during sleep.

    (e) **False** This increases, allowing growth and anabolic repair of tissue wear and tear.

376 **(a)** **True** The median eminence secretes CRH, the releasing hormone for ACTH.

    (b) **False** Aldosterone secretion is regulated mainly by the renin/angiotensin system.

    (c) **True** This negative feedback helps to maintain the blood cortisol level.

    (d) **True** This is part of the circadian rhythm which produces high morning cortisol levels.

    (e) **True** Most forms of stress increase ACTH output by their neural input to the median eminence of the hypothalamus where CRH is formed.

377 **(a)** **False** Melanocyte stimulating hormone is produced in the intermediate lobe of the pituitary; melatonin is produced mainly in the pineal gland.

    (b) **True** The necessary enzymes are in the pineal parenchymal cells.

    (c) **False** It has no role in regulation of human skin pigmentation.

    (d) **True** Melatonin secretion has a pronounced circadian rhythm, low during the day and high by night.

    (e) **True** Melatonin secreted in relation to prevailing conditions of light/darkness may adjust pituitary hormonal rhythms appropriately.

378 **(a)** **False** Increased metabolism leads to peripheral vasodilation.

    (b) **True** The frequency of defaecation increases in hyperthyroidism.

    (c) **True** Thyroid hormones uncouple oxidation from phosphorylation so that more energy appears as heat.

    (d) **False** The reverse is true.

    (e) **True** This suggests a direct action on cells in the sinoatrial node.

379 **(a)** **False** This reduces aldosterone secretion.

    (b) **False** This increases ADH secretion.

    (c) **True** K^+ has a direct stimulatory effect on the adrenal cortex.

    (d) **True** This leads to formation of angiotensin II which stimulates the cortex.

    (e) **True** Though the main action of ACTH is on glucocorticoid secreting cells; it has some action on mineralocorticoid secreting cells.

380 Glucocorticoid injections lead to increases in:
    (a) Lymph gland size
    (b) Fibroblastic activity
    (c) Anabolic activity in muscle
    (d) Bone resorption
    (e) Membrane stability in mast cell and lysosomes

381 An intravenous infusion of noradrenaline differs from one of adrenaline in that it:
    (a) Acts on alpha adrenoceptors
    (b) Does not act on beta adrenoceptors
    (c) Raises total peripheral resistance
    (d) Increases cardiac output
    (e) Decreases skin blood flow

382 Growth hormone:
    (a) Promotes positive nitrogen and phosphorus balance
    (b) Secretion is under hypothalamic control
    (c) Levels in the blood are higher in children than in adults
    (d) Secretion surges during sleep
    (e) Stimulates the liver to secrete somatomedins which regulate bone and cartilage growth

383 Parathormone:
    (a) Secretion is regulated by a pituitary feedback control system
    (b) Acts directly on bone to increase bone resorption
    (c) Decreases the urinary output of calcium
    (d) Decreases phosphate excretion
    (e) Promotes absorption of calcium from the intestines

384 Antidiuretic hormone (vasopressin):
    (a) Is secreted by posterior pituitary gland cells
    (b) Causes the osmolality of the plasma to rise
    (c) Increases the permeability of the cells in the loop of Henle to water
    (d) Secretion is little affected by changes in plasma osmolality of less than 10%
    (e) Secretion increases when plasma volume falls but osmolality is unchanged

380 (a) False Glucocorticoids inhibit mitotic activity in lymphocytes.

(b) False Glucocorticoids inhibit fibroblastic activity; this may allow chronic infections to spread since they are not walled off effectively by fibrous scarring.

(c) False They are catabolic; released amino acids are converted to glucose.

(d) True Decreased bone formation and increased resorption may cause osteoporosis.

(e) True This blocks release of histamine and lysosomal enzymes in allergic responses.

381 (a) False Both act on alpha receptors but noradrenaline is the more potent stimulant.

(b) False Both act on beta receptors but adrenaline is the more potent stimulant.

(c) True Noradrenaline raises but adrenaline reduces it.

(d) False Adrenaline raises but noradrenaline reflexly reduces it.

(e) False Both constrict skin vessels due to their alpha receptor stimulant properties.

382 (a) True It is an anabolic hormone.

(b) True Secretion in the pituitary is stimulated by growth hormone releasing factor and inhibited by somatostatin from the hypothalamus.

(c) False Blood levels are similar in children and adults.

(d) True Sleep is a time for anabolic activity.

(e) True Somatomedins (insulin-like growth factors, IGF) from the liver inhibit the pituitary secretion of growth hormone and stimulate release of somatostatin from the hypothalamus.

383 (a) False It is regulated directly by the calcium level in the blood that perfuses it.

(b) True It stimulates osteoclasts to resorb bone; excessive secretion causes cysts to form.

(c) False The high blood calcium levels with parathormone and the resulting increase in calcium filtration in the glomeruli result in an increased calcium output in urine.

(d) False It increases phosphate excretion by reducing renal phosphate reabsorption.

(e) True It does this indirectly by stimulating 1,25-dihydroxycholecalciferol production.

384 (a) False It is secreted by ADH secreting neurones whose cell bodies lie in the hypothalamus.

(b) False The water retention it induces makes plasma osmolality fall.

(c) False It increases the permeability of the collecting ducts.

(d) False Secretion is affected by 1% changes in osmolality; the sensitivity of the hypothalamic receptors to osmolar change accounts for the constancy of plasma osmolality.

(e) True Volume changes detected by vascular low-pressure receptors affect ADH secretion.

385 Pancreatic glucagon:
   (a) Is produced by the beta cells of the islets of Langerhans
   (b) Is a polypeptide
   (c) Output is inversely proportional to the blood glucose level
   (d) Is essential for the maintenance of a normal blood glucose level
   (e) Increases the breakdown of liver glycogen

386 The concentration of ionized calcium in plasma is:
   (a) The main regulator of parathormone secretion
   (b) Less than the free ionized calcium concentration in cells
   (c) About 50% of the total plasma calcium concentration
   (d) Reduced when plasma pH rises
   (e) Reduced when the plasma protein level rises

387 Cortisol:
   (a) Is bound in the plasma to an alpha globulin
   (b) Is inactivated in the liver and excreted in the bile
   (c) Injections lead to a rise in arterial pressure
   (d) Inhibits release of ACTH from the anterior pituitary gland
   (e) Is released with a circadian variation which is independent of sleep

388 When secretory activity in the thyroid gland increases:
   (a) It takes up iodide from the blood at a faster rate
   (b) Its follicles enlarge and fill with colloid
   (c) The follicular cells become more columnar
   (d) The follicular cells ingest colloid by endocytosis
   (e) The blood level of thyrotropin (TSH) increases

389 Releasing hormones produced in the hypothalamus:
   (a) Are released into the blood stream mainly in the mammillary nuclei
   (b) Pass down nerve axons to reach the pituitary gland
   (c) May control the output of more than one pituitary hormone
   (d) Regulate the release of thyrotropin
   (e) Regulate the release of oxytocin

390 Adrenaline secretion from the adrenal glands increases the:
   (a) Blood glucose level
   (b) Blood free fatty acid level
   (c) Blood flow to skeletal muscle
   (d) Blood flow to the splanchnic area
   (e) Release of renin in the kidneys

385 **(a)** **False** It is produced by the alpha cells.
  (b) **True** It is quite similar in structure to secretin.
  (c) **True** It normally prevents a serious fall in blood glucose.
  (d) **False** After removal of the pancreas, glucose levels can be maintained by insulin only.
  (e) **True** It also mobilizes fatty acids.

386 **(a)** **True** Parathormone secretion is stimulated by a fall in the ionized Ca^{2+} level.
  (b) **False** It is more than 10^5 times higher.
  (c) **True** Most of the rest is bound to protein, mainly albumin.
  (d) **True** In alkalosis more calcium is protein bound.
  (e) **False** The ionized calcium level is regulated independently.

387 **(a)** **True** It is bound to transcortin; free cortisol is released to replace that taken up by the tissues.
  (b) **False** The inactive products of cortisol degradation in the liver are conjugated with glucuronic acid and sulphate and excreted in the urine.
  (c) **True** Partly at least because of its mineralocorticoid effects.
  (d) **True** The negative feedback loop that maintains plasma cortisol levels constant.
  (e) **True** It is regulated through a hypothalamic 'clock'.

388 **(a)** **True** Iodide uptake is an index of activity.
  (b) **False** The follicles shrink as the colloid content falls.
  (c) **True** They change from cuboidal to columnar as their activity increases.
  (d) **True** Reabsorption lacunae form as thyroglobulin is broken down to release hormones.
  (e) **False** Negative feedback causes TSH levels to fall.

389 **(a)** **False** They are released mainly in the median eminence.
  (b) **False** They are carried in the hypophyseal-portal vessels from median eminence to anterior pituitary gland.
  (c) **True** For example, gonadotrophin-releasing hormone (GnRH) controls pituitary secretion of follicle-stimulating hormone (FSH) and luteinizing hormone (LH).
  (d) **True** Thyrotropin-releasing hormone (TRH) controls thyrotropin release.
  (e) **False** Oxytocin is released from terminals of nerves running from hypothalamus to posterior pituitary gland.

390 **(a)** **True** By promoting glycogenolysis in the liver.
  (b) **True** By promoting lipolysis in the fat stores.
  (c) **True** By its predominant effect on beta receptors in arteriolar smooth muscle.
  (d) **False** Splanchnic flow falls since alpha receptors predominate in splanchnic arterioles.
  (e) **True** Juxtaglomerular cells respond to beta receptor stimulation by releasing renin.

391 Thyroid-stimulating hormone (TSH) is secreted at a higher than normal rate:
- (a) After partial removal of the thyroid gland
- (b) In infants born without a thyroid gland
- (c) When metabolic rate falls
- (d) In starvation
- (e) When the diet is deficient in iodine

392 Insulin:
- (a) Stimulates release of free fatty acids from adipose tissue
- (b) Secretion tends to raise the plasma potassium level
- (c) Facilitates entry of glucose into skeletal muscle
- (d) Facilitates entry of amino acids into skeletal muscle
- (e) Secretion is increased by vagal nerve activity

393 The pituitary gland:
- (a) Regulates activity in all other endocrine glands
- (b) Secretes prolactin when stimulated by a hypothalamic releasing factor
- (c) Secretes antidiuretic hormone when blood osmolality falls
- (d) Has an intermediate lobe which secretes melanotropin
- (e) Responds to nervous and hormonal influences from the brain

394 Thyrocalcitonin:
- (a) Is produced by thyroid follicular cells
- (b) Increases basal metabolic rate
- (c) Reduces blood calcium in parathyroidectomized animals
- (d) Secretion occurs when the blood phosphate level rises
- (e) Stimulates osteoclast activity

395 The thyroid gland:
- (a) Takes up iodide against its electrochemical gradient
- (b) Decreases in size when dietary iodine is deficient
- (c) Is relatively avascular
- (d) Contains enzymes which oxidize iodide to iodine
- (e) Contains enzymes which iodinate tyrosine

396 Adrenaline differs from noradrenaline in that it:
- (a) Increases the heart rate when injected intravenously
- (b) Is the main catecholamine secreted by the adrenal medulla
- (c) Increases the strength of myocardial contraction
- (d) Is a more potent dilator of the bronchi
- (e) Constricts blood vessels in mucous membranes

391 (a) True Due to a reduction in pituitary inhibition by circulating thyroxine.
   (b) True Due to absence of the normal pituitary inhibition by circulating thyroxine.
   (c) False TSH and thyroxine influence metabolic rate, not vice versa.
   (d) False The level falls, which tends to conserve energy.
   (e) True Due to inadequate manufacture of thyroxine, pituitary inhibition is reduced.

392 (a) False It stimulates uptake of fatty acids by adipose tissue.
   (b) False It lowers it by promoting potassium uptake by cells.
   (c) True Thus lowering the blood sugar level.
   (d) True Thereby favouring anabolism.
   (e) True This mobilizes insulin at the beginning of a meal.

393 (a) False For example, parathyroid activity is not regulated by the pituitary.
   (b) True From the anterior pituitary when stimulated by prolactin-releasing hormone (PRH).
   (c) False ADH is secreted by the posterior pituitary in response to a rise in blood osmolality.
   (d) True This may stimulate melanin production in human melanocytes.
   (e) True The anterior pituitary is influenced by hormones arriving in portal-hypophyseal vessels and the posterior pituitary by impulses travelling in the hypothalamo-hypophyseal tract.

394 (a) False It is produced by parafollicular cells.
   (b) False It does not affect basal metabolic rate.
   (c) True Its action is independent of the parathyroid glands.
   (d) False It is released when the blood calcium level rises.
   (e) False It stimulates bone deposition by osteoblasts.

395 (a) True It is taken by an active process (the iodide pump).
   (b) False Hyperplasia due to TSH stimulation occurs to give goitre.
   (c) False It has one of the highest blood flow rates in the body; bleeding during surgery may be a problem.
   (d) True This is a stage in the formation of thyroxine.
   (e) True Iodination takes place in the colloid.

396 (a) True Noradrenaline injection causes reflex slowing of the heart.
   (b) True Adrenaline constitutes some 80% of this secretion.
   (c) False Both increase the strength of myocardial contraction.
   (d) True It has stronger beta effects (including bronchodilation).
   (e) False Both vasoconstrict ('decongest') mucous membranes.

397 Growth hormone secretion:
  (a) Is stimulated by somatostatin released from the hypothalamus
  (b) Increases when the blood glucose level falls
  (c) Has a lactogenic effect
  (d) Increases the size of viscera
  (e) Stimulates liver production of somatomedins

398 Vitamin D:
  (a) Increases the intestinal absorption of calcium
  (b) Is essential for normal calcification in childhood
  (c) Requires hepatic modification for activation
  (d) Cannot be formed in the body
  (e) Deficiency may result in hyperparathyroidism

399 Prolactin:
  (a) Has a similar chemical structure and physiological action to luteinizing
      hormone
  (b) Is responsible for breast growth in puberty
  (c) Release is inhibited by dopamine
  (d) Secretion is stimulated by suckling of the breast
  (e) Causes preformed milk to be ejected by the breast during suckling

400 The level of ionized calcium in blood falls when:
  (a) Blood phosphate levels fall
  (b) Subjects hyperventilate
  (c) The thyroid gland is perfused with a calcium-rich solution
  (d) Plasma protein levels fall
  (e) Sodium citrate is added

401 Thyroxine:
  (a) Is stored in the follicular cells as thyroglobulin
  (b) Increases the resting rate of carbon dioxide production
  (c) Is essential for normal development of the brain
  (d) Is essential for normal red cell production
  (e) Acts more rapidly than triiodothyronine (T3)

402 Parathormone:
  (a) Decreases the renal clearance of phosphate
  (b) Mobilizes bone calcium independently of its actions on the kidney
  (c) Depresses the activity of the anterior pituitary gland
  (d) In the blood rises when the calcium level falls
  (e) Stimulates the final activation of vitamin D (cholecalciferol) in the kidney

403 The chemical structure of insulin:
  (a) Is made up of three peptide chains
  (b) Is identical in all mammalian species
  (c) Is such that it is effective when taken by mouth
  (d) Has been synthesized in the laboratory
  (e) Can be synthesized by bacteria

397 **(a) False** Somatostatin inhibits growth hormone secretion.
    (b) True This is the basis of a test of pituitary function.
    (c) True Prolactin and growth hormone are similar peptides.
    (d) True It stimulates growth of most tissues.
    (e) True These peptides mediate general stimulation of growth.

398 **(a) True** This occurs mainly in the upper small intestine.
    (b) True In its absence bones are weak and deformed (rickets).
    (c) True Initial (25-)hydroxylation occurs here.
    (d) False It is formed by sunlight from ergosterol in the skin.
    (e) True The low blood calcium level stimulates parathormone secretion.

399 **(a) False** They are distinct hormones with different actions.
    (b) False Breast growth depends on oestrogens and progestogens.
    (c) True This mediates hypothalamic regulation of prolactin levels.
    (d) True This is responsible for the maintenance of lactation in the puerperium.
    (e) False Milk ejection results from oxytocin secretion.

400 **(a) False** It rises since the product of $[Ca^+]$ $[PO_4^-]$ is constant.
    (b) True Alkalosis increases calcium binding by plasma proteins.
    (c) True Due to release of thyrocalcitonin.
    (d) False But the total calcium level falls.
    (e) True This binds calcium ions and prevents clotting.

401 **(a) False** It is stored as thyroglobulin in the follicles.
    (b) True By increasing the basal metabolic rate.
    (c) True Deficiency in infancy causes mental retardation (cretinism).
    (d) True Deficiency causes anaemia.
    (e) False T3 acts within a day, thyroxine within 2–3 days.

402 **(a) False** It increases it by depressing phosphate reabsorption.
    (b) True It does so in the absence of the kidneys.
    (c) False It does not affect the anterior pituitary.
    (d) True Blood calcium level determines its rate of secretion.
    (e) True From 25-hydroxy- to 1,25-dihydroxycholecalciferol.

403 **(a) False** There are only two peptide chains.
    (b) False Minor differences occur but do not affect insulin action.
    (c) False Its peptide structure is broken down in the gut.
    (d) True In 1964 by Katsoyannis.
    (e) True Using recombinant DNA.

404 Hormones secreted by the adrenal cortex:
   (a) Include cholesterol
   (b) Are mostly bound to plasma proteins
   (c) Include sex hormones
   (d) Are excreted mainly in the bile after conjugation
   (e) Are essential for the maintenance of life

405 During an oral glucose tolerance test the:
   (a) Subject is given 5–10 g of glucose
   (b) Plasma glucose should rise by less than 10% from the fasting level
   (c) Plasma insulin should rise by about 100% from the fasting level
   (d) Rise in plasma glucose is less than with intravenous administration
   (e) Rise in plasma insulin is less than with intravenous administration

406 Secretin differs from cholecystokinin-pancreozymin (CCK-PZ) in that it:
   (a) Is formed by mucosal cells in the upper small intestine
   (b) Stimulates the pancreas to secrete a juice which is rich in digestive enzymes
   (c) Stimulates the pancreas to secrete a watery alkaline juice
   (d) Has less effect on gall bladder smooth muscle
   (e) Decreases gastric motility

407 Inhibition of angiotensin converting enzyme (ACE) decreases the:
   (a) Formation of angiotensin II
   (b) Plasma renin level
   (c) Work of the heart
   (d) Circulating level of angiotensin I
   (e) Total body potassium

408 The plasma level of adrenocorticotrophic hormone (ACTH):
   (a) Is normally maximal around midnight
   (b) Is regulated mainly by the cortisol level
   (c) Shows exaggerated circadian fluctuations with an adrenal tumour
   (d) Is raised in the presence of complete adrenal failure
   (e) Is raised in patients on long-term high-dosage glucocorticoids

409 Possible consequences of hypothyroidism include having:
   (a) A subnormal mouth temperature
   (b) A tendency to fall asleep frequently
   (c) Increased body hair (hirsutism)
   (d) Moist hands and feet
   (e) Prominent eye balls

410 Sudden complete loss of parathyroid function:
   (a) Leads to skeletal muscle spasms
   (b) May be fatal in the absence of therapy to raise the level of ionized calcium
   (c) Causes haemorrhagic disease due to lack of calcium for haemostasis
   (d) May be treated short term by slow intravenous injection of calcium ions
   (e) May be treated long term by regular doses of vitamin D

404 (a) False This is not a hormone.
(b) True For example, the globulin transcortin binds cortisol.
(c) True In both sexes they stimulate axillary and pubic hair.
(d) False After conjugation they are excreted mainly by the kidney.
(e) True Without replacement therapy, loss of adrenal cortical function results in death.

405 (a) False 50–100 g of glucose are used.
(b) False It normally rises by around 50%.
(c) False It rises about ten-fold from a very low fasting level.
(d) True The rise is about half as great.
(e) False Oral glucose stimulates much more release of insulin.

406 (a) False Both are formed by these cells.
(b) False Secretin juice is poor in enzymes.
(c) True Secretin juice is copious and rich in bicarbonate ions.
(d) True 'Cholecystokinin' implies stimulation of the gall bladder.
(e) True CCK-PZ resembles gastrin and increases gastric motility.

407 (a) True The enzyme converts angiotensin I into angiotensin II.
(b) False It rises as the blood pressure falls.
(c) True It is an effective treatment for heart failure.
(d) False It rises due to the increased renin and the inability to convert to angiotensin II.
(e) False Due to the fall in aldosterone secretion, less potassium is excreted.

408 (a) False It is maximal around the time of awakening.
(b) False This is over-ridden by the hypothalamic circadian rhythm.
(c) False A low level is maintained by negative feedback.
(d) True Due to loss of negative feedback by cortisol.
(e) False ACTH is suppressed by these exogenous glucocorticoids.

409 (a) True Due to the lowered metabolic rate.
(b) True Due to the slowing of mental processes.
(c) False Hair loss is characteristic of hypothyroidism.
(d) False Decreased sweating is found in hypothyroidism.
(e) False Prominent eye balls are characteristic of the exophthalmos of hyperthyroidism.

410 (a) True This is a central feature of tetany.
(b) True Due to severe convulsions.
(c) False Calcium levels do not fall below the levels needed for haemostasis.
(d) True This is the acute treatment of choice, e.g. calcium gluconate.
(e) True This acts by increasing intestinal calcium absorption.

411 When a patient with diabetes insipidus is treated successfully with antidiuretic hormone, the:
  (a) Urinary flow rate should fall by about 50%
  (b) Urinary output should be reduced to around 5 ml/min
  (c) Urinary osmolality should rise to between 100 and 200 mosmol/litre
  (d) Salt intake should be carefully regulated
  (e) Blood pressure should stabilize within the normal range

412 Severe uncontrolled diabetes mellitus leads to a raised:
  (a) H^+ ion concentration in body fluids
  (b) Plasma K^+ concentration
  (c) Urinary specific gravity and osmolality
  (d) Blood volume
  (e) Arterial P_{CO_2}

413 Hyperthyroidism is associated with a:
  (a) Positive nitrogen balance
  (b) Decreased urinary excretion of calcium
  (c) Clinical picture consistent with excessive beta adrenoceptor stimulation
  (d) Diminished heat tolerance
  (e) Rise in the level of thyroxine-binding protein in plasma

414 An adrenal medullary tumour (phaeochromocytoma) may cause an increase in:
  (a) Systolic blood pressure which may be transient or constant
  (b) Tremor of the extended hand
  (c) Basal metabolic rate
  (d) Diastolic arterial pressure which does not respond to alpha adrenoceptor-blocking drugs
  (e) Urinary catecholamines

415 Short stature is seen in adults who in childhood suffered from:
  (a) Chronic malnutrition
  (b) Castration
  (c) Premature puberty
  (d) Thyroid deficiency
  (e) Adrenal deficiency

416 Insulin:
  (a) Requirements at night are similar to those during the day
  (b) Half-life is usually reduced in patients with diabetes mellitus
  (c) Is partly bound to larger proteins in the blood
  (d) Requirements are increased in obesity
  (e) Requirements are increased by exercise

411 **(a) False** Typically it will be reduced by about 80%.
    (b) False It should fall to the normal value of about 1 ml/min.
    (c) False It should rise to 2–3 times normal plasma osmolality (600–900 mosmol/litre).
    (d) False ADH does not interfere with salt regulation.
    (e) True Due to greater stability of the body fluids.

412 **(a) True** This is a prime feature of ketoacidosis.
    (b) True The excess H^+ ions compete with K^+ ions for excretion in the distal tubules.
    (c) True Due to the dissolved glucose.
    (d) False This falls due to osmotic diuresis and vomiting.
    (e) False Hyperventilation reduces P_{CO_2} to compensate the metabolic acidosis.

413 **(a) False** It is negative due to muscle wasting.
    (b) False It rises due to liberation of calcium from bone.
    (c) True Beta adrenoceptor-blocking drugs relieve such features, e.g. tachycardia.
    (d) True Heat intolerance is due to increased heat production.
    (e) False The protein levels are normal but they bind more thyroxine.

414 **(a) True** Due to phasic or tonic release of adrenaline and/or noradrenaline.
    (b) True Due to beta adrenoceptor stimulation by adrenaline.
    (c) True Due to release of adrenaline.
    (d) False α-receptor blockers typically lower the blood pressure.
    (e) True This is a diagnostic feature.

415 **(a) True** Early stunting cannot be compensated for later in childhood.
    (b) False This leads to increased height due to delayed closure of the epiphyses.
    (c) True The sex hormones promote early closure of the epiphyses.
    (d) True Thyroid hormones are essential for normal growth.
    (e) True Adrenal hormones also are essential for normal growth.

416 **(a) False** Insulin is required mainly in response to meals.
    (b) False The disease is not usually due to rapid insulin breakdown.
    (c) True Abnormal binding may occur in diabetes mellitus.
    (d) True Obese patients usually show increased insulin resistance.
    (e) False Exercise reduces insulin requirements.

417 The risk of tetany is increased by:
  (a) Sudden rises in plasma bicarbonate
  (b) Sudden rises in plasma magnesium
  (c) Removal of the anterior pituitary gland
  (d) The onset of respiratory failure
  (e) The onset of renal failure

418 Destruction of the anterior pituitary (Simmond's disease) causes:
  (a) Amenorrhoea
  (b) Diabetes insipidus
  (c) Failure of ovulation
  (d) Impaired ability to survive severe stress
  (e) A rise in basal metabolic rate

419 Removal of the thyroid gland (without replacement therapy) leads to an increased:
  (a) Blood TSH level
  (b) Blood cholesterol level
  (c) Blood glucose level during an oral glucose tolerance test
  (d) Response time for tendon reflexes
  (e) Tremor of the fingers

420 In severe diabetes mellitus, there may be a fall in:
  (a) Extracellular fluid osmolality
  (b) Appetite
  (c) Blood volume
  (d) Arterial blood pH to below 7.0
  (e) Blood bicarbonate to half its normal value

421 Excessive glucocorticoid production (Cushing's syndrome) causes an increase in:
  (a) Skin thickness
  (b) Bone strength
  (c) Blood glucose
  (d) Arterial pressure
  (e) The rate of wound healing

422 A pituitary tumour secreting excess growth hormone in an adult may lead to:
  (a) A homonymous hemianopia
  (b) Giantism
  (c) Reduced levels of somatomedins in blood
  (d) Enlargement of the liver
  (e) A raised blood glucose level

417 (a) True In alkalosis, the calcium-binding power of the plasma proteins increases.
   (b) False Like calcium, magnesium ions tend to prevent tetany.
   (c) False The pituitary is not involved in calcium homeostasis.
   (d) False The acidosis in respiratory failure reduces calcium binding by protein.
   (e) False The acidosis in renal failure also reduces calcium binding by protein.

418 (a) True Due to absence of FSH and LH.
   (b) False ADH is released from the posterior pituitary.
   (c) True Due to absence of FSH and LH; infertility results.
   (d) True Due to loss of ACTH and failure of the cortisol surge in response to stress; loss of TSH and consequent hypothyroidism also contribute.
   (e) False BMR falls due to loss of TSH drive to the thyroid.

419 (a) True Due to loss of negative feedback to the pituitary.
   (b) True Due mainly to a reduction in cholesterol excretion.
   (c) False The curve is flattened due to slow absorption of glucose.
   (d) True The 'hung up' ankle jerk is a good example.
   (e) False Tremor is a feature of hyperthyroidism.

420 (a) False It rises due to excess glucose molecules plus water loss.
   (b) True Due to nausea and impaired consciousness.
   (c) True Due to osmotic diuresis and vomiting.
   (d) True This indicates life-threatening acidosis.
   (e) True Bicarbonate is used up buffering the ketoacids.

421 (a) False Skin is thin due to protein catabolism; striae appear.
   (b) False Bones are weakened by breakdown of the protein matrix.
   (c) True This is a major effect of glucocorticoids.
   (d) True Due to the salt and water retention caused by glucocorticoids.
   (e) False Healing is slowed in this catabolic state.

422 (a) False Damage to the crossing nasal retinal fibres in the optic chiasma leads to bitemporal hemianopia.
   (b) False After puberty when the epiphyses have closed, excess GH causes acromegaly.
   (c) False GH leads to increased production of somatomedins in the liver.
   (d) True Body organs as well as the peripheries increase in size in acromegaly.
   (e) True Growth hormone has 'diabetogenic' effects.

423 Hypoglycaemic coma differs from hyperglycaemic coma in that there is more likelihood of:
  (a) A rapid loss of consciousness
  (b) A weak pulse
  (c) Normal blood pH
  (d) Glucose-free urine
  (e) A high acetone level in urine

424 In adrenal failure there is likely to be a fall in the:
  (a) Extracellular fluid volume
  (b) Total red cell mass
  (c) Sodium:potassium ratio in plasma
  (d) Arterial blood pressure
  (e) Blood urea

425 In diabetic ketosis there is a decreased metabolic breakdown of:
  (a) Ketones
  (b) Glycogen
  (c) Glucose
  (d) Fat
  (e) Amino acids

426 A patient with severe diabetic ketoacidosis is likely to benefit from administration of:
  (a) Intragastric fluids
  (b) Intravenous insulin
  (c) Isotonic glucose
  (d) Isotonic sodium chloride
  (e) Oxygen by breathing mask if hyperventilation is present

427 Impaired growth hormone secretion:
  (a) In children causes delayed puberty
  (b) In children leads to short stature with more stunting of the limbs than the trunk
  (c) Is associated with pale, fine and soft skin
  (d) In adults leads to a reduction in the size of the viscera
  (e) Can be treated effectively with bovine growth hormone

428 Parathormone secretion is usually increased:
  (a) In patients with chronic renal failure
  (b) In people taking excessive amounts of vitamin D
  (c) In patients with anterior pituitary tumours secreting excessive amounts of its hormones
  (d) When blood phosphate levels fall
  (e) When plasma protein levels fall

423 (a) True Blood glucose can drop more rapidly than diabetic ketosis can develop.

(b) False The pulse is usually strong in hypoglycaemic coma but weak in hyperglycaemic coma because of fluid depletion.

(c) True Hypoglycaemia does not change the pH.

(d) True However, glucose may be present if urine containing glucose entered the bladder before the onset of hypoglycaemia.

(e) False Usually acetone is absent in hypoglycaemic coma.

424 (a) True Due to salt and water loss from lack of glucocorticoids and mineralocorticoids.

(b) True But the haemoglobin level rises due to haemoconcentration.

(c) False It falls; loss of aldosterone leads to retention of potassium.

(d) True Low blood volume may lead to hypotension and hypovolaemic circulatory failure.

(e) False It tends to rise due to the oliguria associated with the hypotension.

425 (a) False Breakdown continues normally but ketones accumulate due to rapid production.

(b) False Insulin normally inhibits glycogenolysis.

(c) True Due to its impaired entry into the cells.

(d) False Fat breakdown is increased to yield ketone bodies.

(e) False Gluconeogenesis and amino acid catabolism increase.

426 (a) False Vomiting is likely so intravenous fluids are needed to correct the fluid deficit.

(b) True Insulin is needed to reverse the derangement of metabolism.

(c) True A water deficit is remedied by intravenous isotonic glucose.

(d) True This remedies the extracellular fluid deficit; the pH disturbance is corrected by restoring normal metabolism and fluid balance.

(e) False The hyperventilation is due to acidosis, not oxygen lack.

427 (a) False Pituitary gonadotrophins determine the onset of puberty.

(b) False Pituitary dwarfs are usually normally proportioned.

(c) True In addition, body hair is normally sparse.

(d) False It has no detectable effect on organ size in adults.

(e) False Only the human form is effective.

428 (a) True Phosphate retention results in a fall in the ionized calcium level in blood; this stimulates the parathyroid to produce more parathormone (secondary hyperparathyroidism).

(b) False The increased level of ionized calcium in blood depresses parathyroid activity.

(c) False Pituitary hormones are not involved in the regulation of parathyroid activity.

(d) False This raises ionized calcium levels and depresses parathyroid activity.

(e) False This decreases the total blood calcium but not the ionized calcium level.

429 An oral glucose tolerance test in a patient with:
   (a) Diabetes mellitus shows a similar fasting blood glucose level to that in a normal person
   (b) Diabetes mellitus shows glycosuria when blood glucose is three times the normal fasting level
   (c) Diabetes mellitus shows a delayed return to the fasting blood glucose level
   (d) An insulin-secreting tumour shows no rise in blood glucose level during the test
   (e) Malabsorption syndrome shows a lower than normal peak level for blood glucose

430 Surgical removal of the pituitary gland is likely to lead to a decrease in:
   (a) Plasma osmolality
   (b) Menstrual frequency
   (c) Axillary hair
   (d) Sexual desire (libido)
   (e) Breast size

429 **(a) False**   The level is higher due to impaired glucose homeostasis even in the fasting state.

    (b) True   The renal threshold for glucose is about twice the normal fasting level.

    (c) True   Due to impaired insulin response to the glucose stimulus.

    (d) False   Blood glucose rises but then falls to a low level due to excessive insulin secretion.

    (e) True   The curve is flattened due to impaired glucose absorption.

430 **(a) False**   Osmolality increases due to the induced diabetes insipidus.

    (b) True   Amenorrhoea is common due to loss of gonadotrophic hormones.

    (c) True   The adrenal androgens responsible for axillary hair are under ACTH control.

    (d) True   Libido is influenced by the sex hormones which are under gonadotrophic control.

    (e) True   The oestrogen and progesterone responsible for breast development are under gonadotrophic control.

9 REPRODUCTIVE SYSTEM

431 In the normal menstrual cycle:
- (a) Blood loss during menstruation averages around 100 ml
- (b) The proliferative phase depends on oestrogen secretion
- (c) Cervical mucus becomes more fluid around the time of ovulation
- (d) Ovulation is followed by a surge in blood luteinizing hormone level
- (e) Basal body temperature is higher after ovulation

432 Fertilization of the human ovum normally:
- (a) Occurs in the uterus
- (b) Prevents further spermatozoa from entering the ovum
- (c) Occurs 2–5 days after ovulation
- (d) Occurs 5–7 days before implantation
- (e) Leads to the secretion of human chorionic gonadotrophin (HCG) within 2 weeks

433 Human spermatozoa:
- (a) Contain 23 chromosomes
- (b) Have enzymes in their heads which aid penetration of the ovum
- (c) Are produced faster at 37°C than at 32°C
- (d) Are motile in the seminiferous tubules
- (e) Are stored mainly in the seminal vesicles

434 After a baby is born, there is normally a fall in:
- (a) Its systemic vascular resistance
- (b) Its pulmonary vascular resistance
- (c) Direct flow from pulmonary artery to aorta
- (d) Direct flow from right to left atrium
- (e) Direct flow from right to left ventricle

435 Secretion of testosterone:
- (a) Depresses pituitary secretion of LH
- (b) Causes the epiphyses of long bones to unite
- (c) May lead to a negative nitrogen balance
- (d) Stimulates growth of scalp hair
- (e) Stimulates growth of body hair

436 Human chorionic gonadotrophic hormone (HCG):
- (a) Is a steroid
- (b) Acts directly on the uterus to maintain the endometrium
- (c) Is formed in the anterior pituitary
- (d) Blood level rises steadily throughout pregnancy
- (e) Can be detected in the urine as an early sign of pregnancy

437 Compared with the adult, the newborn has less ability to:
- (a) Excrete bilirubin
- (b) Maintain a constant body temperature
- (c) Tolerate brain hypoxia
- (d) Manufacture antibodies
- (e) Resist infection

431 **(a) False** It varies widely but averages about 30 ml (one ounce).
    (b) True This occurs in the first half of the cycle.
    (c) True This mucus 'cascade' may facilitate sperm passage.
    (d) False The LH surge precedes and initiates ovulation.
    (e) True Metabolic rate is raised by progesterone.

432 **(a) False** It occurs in the outer third of the uterine tube.
    (b) True The zona pellucida becomes impermeable.
    (c) False The ovum remains viable for only about a day.
    (d) True During this time the fertilized ovum travels along the uterine tube and spends several days free in the uterus.
    (e) True This is necessary to maintain ovarian hormone production whose withdrawal causes endometrial necrosis.

433 **(a) True** Half the complement of human somatic cells.
    (b) True These are in the acrosome ('extremity body').
    (c) False Normal core temperature inhibits formation of spermatozoa.
    (d) False At this stage they are non-motile and cannot fertilize.
    (e) False They are stored in the epididymis.

434 **(a) False** This rises due to closure of the umbilical arteries.
    (b) True Due to expansion of the lungs and their blood vessels.
    (c) True Flow in the ductus arteriosus reverses due to (a) and (b).
    (d) True Again due to reversal of the pressure gradient.
    (e) False Normally there is no opening in the intraventricular septum.

435 **(a) True** This negative feedback keeps blood testosterone constant.
    (b) True Sexual precocity can cause short stature.
    (c) False Testosterone is anabolic and leads to skeletal muscle hypertrophy.
    (d) False Scalp hair tends to recede.
    (e) True This is a male secondary sexual characteristic.

436 **(a) False** It is a glycoprotein resembling luteinizing hormone (LH).
    (b) False It acts on the ovaries to maintain the corpus luteum.
    (c) False It is formed in the chorion of the developing embryo.
    (d) False It peaks in the first 3 months of pregnancy and then declines.
    (e) True An immunological technique is used to identify it.

437 **(a) True** 'Physiological' jaundice is due to immaturity of the liver.
    (b) True Temperature-regulating mechanisms are also immature.
    (c) False Fetal tissues are adapted to relative hypoxia.
    (d) True Immunological competence develops around 3 months of age.
    (e) False Maternal antibodies (supplied via the placenta and breast) provide effective passive immunity.

438 Androgens are secreted in the:
   (a) Testis by the seminiferous tubules
   (b) Fetus in greater quantities than in early childhood
   (c) Female by the ovary
   (d) Male by the adrenal cortex
   (e) Male in decreasing quantities after the age of 30

439 During pregnancy the:
   (a) Uterine muscle enlarges due mainly to cell proliferation
   (b) Uterus is quiescent until the onset of labour
   (c) Breasts enlarge due mainly to the action of prolactin
   (d) Haematocrit rises
   (e) Basal metabolic rate rises by more than 10%

440 Males differ from females in that their:
   (a) Pituitary glands secrete different gonadotrophic hormones
   (b) Hypothalamus shows different patterns of hormone secretion
   (c) Gonads produce gametes until later in life
   (d) Blood gonadotrophin levels do not rise in later life
   (e) Polymorphs show 'drumsticks' of chromatin on their nuclei

441 Fetal haemoglobin:
   (a) Is the only type identifiable in fetal blood
   (b) Forms the bulk of total haemoglobin for the first year of life
   (c) Has a higher oxygen-carrying capacity than adult haemoglobin
   (d) Binds 2,3-DPG more avidly than does adult haemoglobin
   (e) Has a higher affinity than the adult form for oxygen at low P_{O_2}

442 For normal development and fertility of spermatozoa there must be:
   (a) Secretion of testosterone
   (b) Secretion of luteinizing hormone
   (c) Secretion of follicle-stimulating hormone
   (d) A testicular temperature of 37°C
   (e) A sperm count of more than 10^{10}/ml

443 In the mammary glands:
   (a) Milk formation is stimulated by oestrogen and progesterone
   (b) Milk formation can be depressed by hypothalamic activity
   (c) Maintenance of lactation depends on suckling
   (d) Lactation ceases if the anterior pituitary gland is destroyed
   (e) Milk ejection ceases if the posterior pituitary gland is destroyed

444 The male postpubertal state differs from the prepubertal in that:
   (a) The gonads are responsive to gonadotrophic hormones
   (b) There is a greater output of 17-ketosteroids in the urine
   (c) Skeletal muscle is stronger per unit mass of tissue
   (d) The circulating level of follicle-stimulating hormone is higher
   (e) Hypothalamic output of gonadotrophin-releasing factors is greater

438 (a) **False** They are secreted by the interstitial cells of the testis
   (b) **True** Fetal androgens control male sex organ development.
   (c) **True** The ovaries secrete small amounts of androgen.
   (d) **True** Adrenal androgens control pubic and axillary hair growth in both males and females.
   (e) **True** There is a gradual fall, with no clear-cut 'climacteric'.

439 (a) **False** The enlargement is due more to an increase in the size of the muscle cells.
   (b) **False** Spontaneous uterine contractions occur during pregnancy.
   (c) **False** The enlargement is due mainly to oestrogen and progesterone.
   (d) **False** It falls due to the increase in plasma volume.
   (e) **True** It increases by about one-third.

440 (a) **False** The gonadotrophic hormones are the same in males and females.
   (b) **True** The female monthly cycle originates here.
   (c) **True** Much later than the female menopausal age.
   (d) **False** Both sexes show a rise in FSH and LH levels.
   (e) **False** It is the female polymorphs that show this.

441 (a) **False** The adult type appears around mid-gestation.
   (b) **False** It has almost disappeared by 4 months.
   (c) **False** They have similar oxygen capacities.
   (d) **False** It binds it less readily.
   (e) **True** This aids oxygen transfer in the placenta.

442 (a) **True** In its absence spermatogenesis is depressed.
   (b) **True** This controls secretion of testosterone.
   (c) **True** This also is required by the germinal epithelium.
   (d) **False** Testicular temperature should be maintained around 32°C.
   (e) **False** The normal value is around 10^8/ml.

443 (a) **False** These depress milk formation during pregnancy; prolactin stimulates milk formation.
   (b) **True** By release of prolactin inhibiting hormone (dopamine).
   (c) **True** This causes prolactin secretion which initiates and maintains lactation after delivery.
   (d) **True** Milk formation ceases due to loss of prolactin.
   (e) **True** Due to loss of oxytocin in response to suckling.

444 (a) **False** They are responsive before puberty as well.
   (b) **True** Due to greater production of sex hormones in the body.
   (c) **True** This is one of the actions of testosterone.
   (d) **True** This stimulates growth and function of the testes.
   (e) **True** This initiates the various changes of puberty.

445 In the placenta:
  (a) Fetal and maternal blood mix freely in the sinusoids
  (b) The P_{O_2} in sinusoidal blood is similar to that in maternal arterial blood
  (c) The barrier to oxygen diffusion is much greater than in alveoli
  (d) Fetal blood in umbilical veins has a P_{O_2} within 10% of that in maternal sinusoids
  (e) Fetal blood becomes more than 50% saturated with oxygen

446 During pregnancy, there is an increase in:
  (a) Maternal blood volume to twice the normal level
  (b) Peripheral resistance
  (c) Venous tone
  (d) Ligament laxity
  (e) Maternal parathormone secretion

447 In the fetal circulation the oxygen content of blood in the:
  (a) Femoral artery is less than that in the brachial artery
  (b) Superior vena cava is higher than that in the inferior vena cava
  (c) Right ventricle is higher than that in the left ventricle
  (d) Pulmonary artery is higher than that in the pulmonary veins
  (e) Cerebral arteries is lower than in the maternal cerebral arteries

448 Normal parturition depends on:
  (a) An abrupt fall in placental secretion of oestrogen and progesterone
  (b) Release of oxytocin from the posterior pituitary gland
  (c) Activation of beta adrenoceptors in uterine muscle
  (d) The presence of normal ovaries
  (e) Innervation of the uterus

449 The interstitial cells of the testis:
  (a) Contribute to the volume of seminal fluid
  (b) Are the source of the hormone inhibin
  (c) Are stimulated to secrete by luteinizing hormone (LH)
  (d) Depend on hypothalamic activity to function properly
  (e) Are non-functional unless the testis descends from the abdomen to the scrotum

450 The size of the fetus at birth is likely to be smaller in:
  (a) Small than in large mothers
  (b) Multiple than in single fetus pregnancies
  (c) Smoking than in non-smoking mothers
  (d) Female than in male babies
  (e) Firstborn than in subsequent babies

445 (a) False The two circulations remain discrete.
(b) False It is about half the arterial level in this sluggish flow.
(c) True Due to a much thicker cellular barrier to diffusion.
(d) False It is about 50% lower due to the thick diffusion barrier.
(e) True Due to its high affinity for oxygen.

446 (a) False It increases by about a third.
(b) False It decreases (large uterine vessels, vasodilation).
(c) False Venous tone decreases and varicose veins may develop in the legs.
(d) True Due to relaxin, which helps the birth canal to dilate.
(e) True Large amounts of calcium must be mobilized for the fetus.

447 (a) True Due to deoxygenated pulmonary arterial blood passing through the ductus arteriosus to the descending aorta.
(b) False The IVC receives oxygenated blood from the placenta
(c) False Deoxygenated SVC blood streams to the right ventricle while oxygenated IVC blood streams via the foramen ovale to the left ventricle.
(d) True Since the lungs are not ventilated, oxygen is lost rather than gained in its passage through the fetal lungs.
(e) True Umbilical venous blood is only about 80% saturated with oxygen and fetal arterial oxygen levels cannot exceed this; fetal tissues are adapted to survive in relative hypoxia.

448 (a) False Secretion is maintained at high levels until parturition
(b) False Birth can occur in the absence of the posterior pituitary.
(c) False Beta activation may be used to delay onset of labour.
(d) False Ovaries are not essential in late pregnancy and parturition.
(e) False Parturition can occur with a denervated uterus; the cause of the onset of normal labour is unknown.

449 (a) False They secrete testosterone into the circulation.
(b) False This hormone is produced by the seminiferous tubules.
(c) True LH is the interstitial cell stimulating hormone in the male.
(d) True Gonadotropin-releasing hormone from the hypothalamus is needed for LH secretion.
(e) False Undescended testes can secrete testosterone.

450 (a) True Fetal size has some relationship to uterine capacity.
(b) True Placental capacity per fetus is reduced.
(c) True Excessive alcohol intake during pregnancy also reduces fetal size at birth.
(d) True Males are on average about 200 g heavier.
(e) True On average later children are 200 g heavier.

451 Testosterone secretion from the testis:
  (a) Increases at puberty because LH levels increase
  (b) Is responsible for the growth of facial hair
  (c) Has a negative feedback effect on FSH secretion by the anterior pituitary gland
  (d) Peaks in the early evening
  (e) Is responsible for interest in the opposite sex

452 The corpus luteum:
  (a) Is essential for the secretory phase of the menstrual cycle
  (b) Development is controlled by the pituitary gland
  (c) Secretes hormones in early pregnancy when stimulated by pituitary gland hormones
  (d) Is greyish-white in colour
  (e) Begins to atrophy in the second month of pregnancy

453 Erection of the penis:
  (a) Cannot occur before puberty
  (b) Is normally initiated by venoconstriction
  (c) Depends on adrenergic sympathetic nervous activity
  (d) Cannot occur after cervical spinal cord transection
  (e) Is inhibited by ganglion-blocking drugs

454 The placenta:
  (a) Contains villi which transport glucose into fetal blood
  (b) Can convert glucose into glycogen
  (c) Can store iron and calcium
  (d) Actively transports oxygen into fetal blood
  (e) Allows certain proteins to pass from maternal to fetal blood

455 The ovaries:
  (a) Begin to develop ova at puberty when acted on by FSH
  (b) Are required for cyclical menstrual activity
  (c) Must have double follicular rupture if identical twins are conceived
  (d) Cease to respond to FSH after the menopause
  (e) Secrete hormones which constrict uterine vessels

456 The normal seminal ejaculate:
  (a) Has a volume of about 5–10 ml
  (b) Comes mainly from the seminiferous tubules and epididymis
  (c) Contains fructose from the seminal vesicles
  (d) Contains phosphate and bicarbonate buffers
  (e) Contains prostaglandins

457 The 21st day of the menstrual cycle differs from the seventh in that the:
  (a) Endometrium is thicker and contains glands
  (b) Blood level of progesterone is higher
  (c) Blood is oestrogen free
  (d) Blood level of FSH is at a maximum
  (e) Endometrial glycogen content is higher

451 (a) **True** LH secretion is controlled by hypothalamic GnRH (LH/FSH RH).
  (b) **True** Females given testosterone may develop a beard.
  (c) **False** Inhibin secreted by the testis has a negative feedback effect on FSH secretion.
  (d) **False** It peaks around the time of awakening.
  (e) **True** Castrates have little interest in sex.

452 (a) **True** By its secretion of oestrogen and progesterone.
  (b) **True** It depends on luteinizing hormone (LH).
  (c) **False** It is controlled by human chorionic gonadotrophin (HCG) from the early placenta.
  (d) **False** It is yellow; corpus luteum is Latin for 'yellow body'.
  (e) **False** It is essential for the first 3 months of pregnancy.

453 (a) **False** It is not uncommon in infants.
  (b) **False** It is normally initiated by arteriolar dilation.
  (c) **False** It depends on parasympathetic cholinergic nerves; activity in sympathetic nerves is required for ejaculation.
  (d) **False** It is based on a spinal reflex.
  (e) **True** These block sympathetic and parasympathetic pathways.

454 (a) **True** As in the intestinal villi.
  (b) **True** This glycogen is stored in the placenta.
  (c) **True** Also proteins and fat.
  (d) **False** Oxygen transfer is accounted for by its pressure gradient.
  (e) **True** For example, the gamma globulins concerned with passive immunity.

455 (a) **False** The immature ova are formed before birth and no more are developed after birth.
  (b) **True** Because of their secretion of oestrogen and progesterone.
  (c) **False** Identical twins are derived from a single ovum.
  (d) **True** Follicles disappear and are replaced by fibrous tissue.
  (e) **False** Withdrawal of ovarian hormones leads to vasoconstriction.

456 (a) **False** The normal volume is 2–5 ml.
  (b) **False** These contribute only about 20% of volume.
  (c) **True** Seminal vesicles contribute about 60% of seminal volume (prostate 20%).
  (d) **True** These help to neutralize acidic vaginal fluids.
  (e) **True** Derived from seminal vesicles, prostaglandins may induce contractile activity in the female genital tract.

457 (a) **True** The endometrium is in the secretory phase of the cycle.
  (b) **True** It is very low on the seventh day.
  (c) **False** Oestrogen remains high in the second half of the cycle.
  (d) **False** The maximum is around the 14th day.
  (e) **True** Conditions are optimal for implantation.

458 The newborn baby differs from the adult in that its:
  (a) Urine has a lower maximum osmolality
  (b) Urine has a lower minimum osmolality
  (c) Blood–brain barrier is less permeable to bilirubin
  (d) Temperature regulation is more efficient because of brown fat
  (e) Blood has a greater affinity for oxygen at low oxygen pressures

459 Changes in maternal physiology during pregnancy include a rise in:
  (a) Nitrogen retention
  (b) Mean arterial pressure of around 20 mmHg
  (c) Arterial P_{CO_2}
  (d) Tone in the urinary tract
  (e) The renal threshold for glucose

460 Amniotic fluid is:
  (a) Formed in early pregnancy by filtration from fetal skin capillaries
  (b) Formed in late pregnancy by filtration from the gut mucosa
  (c) Swallowed by the fetus
  (d) Similar in electrolyte composition to plasma
  (e) Inhaled and exhaled by the fetus

461 Ejaculation of semen:
  (a) Depends on a spinal cord reflex
  (b) Depends on sympathetic nerve activity
  (c) Involves rhythmic contractions of striated muscles
  (d) Is accompanied by contraction of the cremasteric muscles
  (e) Is followed by orgasm

462 The fetus normally:
  (a) Gains more weight in the last 10 weeks of gestation than in the first 30 weeks
  (b) Has a higher haemoglobin level at term than a normal adult
  (c) Stores sufficient iron in the liver to last a year after birth
  (d) Has a similar metabolic rate per metre^2 body surface area as an adult
  (e) Passes rectal contents in the last 3 months of gestation

463 Cessation of menstruation (secondary amenorrhoea) may occur because of:
  (a) Psychological stress
  (b) Severe weight loss
  (c) Continuous administration of oestrogens
  (d) An adrenal tumour
  (e) Continuous administration of gonadotropin-releasing hormone (GnRH)

458 (a) **True**   Its ability to concentrate urine is poor.
     (b) **False**   Its ability to dilute urine is also poor.
     (c) **False**   It is more permeable and brain damage can result with high blood bilirubin levels.
     (d) **False**   Temperature regulation is poor in the newborn.
     (e) **True**   Because of persisting fetal haemoglobin.

459 (a) **True**   About 300 g nitrogen is retained, half by maternal tissues and half by fetal tissues.
     (b) **False**   Blood pressure tends to fall, such a rise suggests disease
     (c) **False**   It tends to fall due to increased ventilation.
     (d) **False**   Tone decreases and may lead to ureteric reflux and urinary infections.
     (e) **False**   It falls, and glucose may appear in urine at normal blood glucose levels.

460 (a) **True**   The non-keratinized skin is not waterproof.
     (b) **False**   It is formed by the kidneys and excreted as urine in late pregnancy.
     (c) **True**   Up to 0.5 litre/day is absorbed and excreted as urine.
     (d) **True**   The protein content is, however, much lower.
     (e) **True**   Surfactant from the lungs can be found in amniotic fluid in late pregnancy.

461 (a) **True**   The centre is in the lumbosacral region
     (b) **True**   Stimulation of the hypogastric nerves in man can cause ejaculation.
     (c) **True**   These compress the urethra.
     (d) **True**   This causes elevation of the testicles.
     (e) **False**   Ejaculation and orgasm coincide.

462 (a) **True**   Fetal growth is exponential.
     (b) **True**   Around 170–200 g/litre.
     (c) **False**   The stores last only a few months.
     (d) **False**   It is about twice as great due to rapid growth.
     (e) **False**   Passage of meconium before birth is a sign of distress.

463 (a) **True**   Psychological stress affects hypothalamic activity.
     (b) **True**   Amenorrhoea is common during starvation.
     (c) **True**   Continuous progesterone administration can have the same effect.
     (d) **True**   Androgens from the tumour may oppose the effects of female sex hormones on the endometrium.
     (e) **True**   Normally GnRH secretion is pulsatile; thus interference with the normal hormone rhythms can cause amenorrhoea and infertility.

464 Development of secondary sexual characteristics before age nine could be:
   (a) Due to abnormal secretion of adrenal cortical hormones
   (b) Associated with short stature
   (c) Due to a hypothalamic tumour
   (d) Due to a pituitary tumour
   (e) Present in a normal healthy child

465 The fetal:
   (a) Blood in umbilical veins contains more amino acid than maternal blood in uterine veins
   (b) Aorta has a higher rate of blood flow than the distal pulmonary artery
   (c) Aortic blood pressure is lower than pulmonary arterial pressure
   (d) Systemic resistance is higher than its pulmonary resistance
   (e) Heart rate suggests fetal distress if it exceeds 100 beats/min

466 Lack of pulmonary surfactant:
   (a) Is unlikely in infants born after 30 weeks gestation
   (b) Can be diagnosed by examining the fetal amniotic fluid
   (c) Increases the effort required for expiration
   (d) Decreases the surface tension forces in the lungs
   (e) Leads to poor oxygenation of the blood before birth

467 A child whose sex chromosome pattern is:
   (a) XY develops into a normal female
   (b) XO shows incomplete sexual maturation at puberty
   (c) XXX develops exaggerated female secondary sexual characteristics
   (d) XXY develops into a true hermaphrodite
   (e) XX is less likely to have haemophilia than one with XY

468 The diagnosis of pregnancy is supported by finding:
   (a) Conjugated progesterone in the urine
   (b) Human chorionic gonadotrophin in the urine
   (c) Viscous cervical mucus plugging the cervical canal
   (d) Enlargement of the sebaceous glands in the mammary areolae
   (e) A hardening of the cervical tissue

469 The neonatal:
   (a) Liver stores sufficient vitamin K for the first few months of life
   (b) Blood volume is closer to 750 than 250 ml
   (c) Blood glucose level fluctuates more than the fetal level
   (d) Gut usually lacks certain enzymes needed for digestion of milk
   (e) Peripheral vascular resistance is higher than that of the adult

470 After a child is born:
   (a) Its haemoglobin level rises steadily during the first year
   (b) There should be a delay in clamping the umbilical cord so that blood from the placenta can drain into the fetus
   (c) It should increase its weight by 10% at 4 months
   (d) Its brain can tolerate a lower blood glucose level than that of an adult
   (e) Its brain can tolerate a lower oxygen level than that of an adult

464 (a) True   Adrenal androgens may lead to precocious puberty.
(b) True   Sex hormones cause closure of the epiphyses.
(c) True   If it secretes a gonadotropin-releasing hormone.
(d) True   If gonadotropins are produced.
(e) True   There is a wide scatter in the normal distribution of the age of onset of puberty.

465 (a) True   Amino acids are actively transported from maternal to fetal blood in the placenta.
(b) True   Due to high flow through the ductus arteriosus to the aorta.
(c) True   Blood flows from the pulmonary artery to the distal aorta.
(d) False   Distal aortic flow is greater than distal pulmonary artery flow and the pressure is lower (resistance = pressure/flow).
(e) False   It is normally about 140/min; below 100 suggests distress.

466 (a) False   Its formation starts around the 35th week.
(b) True   Fetal breathing movements wash it into this fluid.
(c) False   It increases the work of inspiration.
(d) False   Without surfactant the surface tension forces are great; these forces must be overcome during inspiration.
(e) False   The lungs are not used for gas exchange before birth; lack of surfactant causes poor oxygenation in the neonate.

467 (a) False   He develops into a normal male.
(b) True   The gonads fail to develop (Turner's syndrome).
(c) False   Abnormalities do not result.
(d) False   He develops as a male with abnormal testes and a high risk of mental retardation (Klinefelter's syndrome).
(e) True   Haemophilia is an X-linked recessive condition.

468 (a) False   This is normally present in the childbearing period.
(b) True   This is not otherwise present.
(c) True   Progesterone increases the viscosity of cervical mucus.
(d) True   These are known as Montgomery's tubercles.
(e) False   The cervix softens during pregnancy.

469 (a) False   Vitamin K is not stored and deficiency at birth is common.
(b) False   It would be about 240 ml in a 3 kg neonate.
(c) True   The neonatal liver is less mature than the maternal liver.
(d) False   It can usually deal adequately with the nutrients in milk.
(e) True   The AV pressure gradient is almost as great as in the adult and flow is much smaller (resistance = pressure gradient/flow).

470 (a) False   It falls from around 170–200 g/litre to around 110 g/litre.
(b) True   The placenta contains about half as much blood as the fetus; some can be transferred by uterine contraction.
(c) False   Its weight should be doubled at this stage and trebled at 1 year.
(d) True   It can tolerate about 25% of the normal adult fasting level.
(e) True   Fetal tissues are adapted to survive moderate hypoxia.

471 Removal of the testes in the adult causes:
    (a) A rise in the pitch of the voice
    (b) Loss of libido
    (c) Loss of the ability to copulate
    (d) Hot flushes, irritability and depression
    (e) A fall in the blood levels of LH and FSH

472 Administration of oestrogens and progestogens to women:
    (a) Prevents menstruation if given daily throughout the year
    (b) Tends to cause salt and water retention
    (c) Depresses secretion of pituitary gonadotrophins
    (d) Decreases the likelihood of ovulation
    (e) Tends to accentuate acne vulgaris

473 Secondary sexual characteristics do not develop in children:
    (a) Who have been castrated
    (b) Whose seminiferous tubules, but not interstitial cells, have been damaged by radiation
    (c) Suffering from severe malnutrition
    (d) With dwarfism
    (e) Lacking pituitary hormones

474 Methods of reducing fertility include:
    (a) Confining intercourse to the period from the 10-20th day of the menstrual cycle
    (b) Bilateral ligation and division of the uterine tubes
    (c) Bilateral ligation and division of the vas deferens
    (d) The use of agents which prevent the fertilized ovum from implanting
    (e) Mechanical barriers (condoms and caps), which are the most effective methods

475 Infertility usually occurs:
    (a) When the sperm count is reduced to 10% of normal
    (b) When posterior pituitary function is lost
    (c) When one uterine tube is blocked
    (d) In males with undescended testes
    (e) Because of a reproductive disorder in the female partner

476 Maternal blood loss in the first 24 h after delivery:
    (a) Is considered abnormal if it exceeds 600 ml
    (b) Is greater after a short than after a long labour
    (c) Is increased if part of the placenta is retained
    (d) May justify transfusion of unmatched AB Rh-positive blood
    (e) May, if excessive, impair anterior pituitary function

471 (a) **False**   Laryngeal changes induced by testosterone are permanent.
   (b) **True**   Testosterone secretion increases libido.
   (c) **False**   The sexual reflexes persist in the absence of testosterone.
   (d) **True**   Rather like menopausal symptoms.
   (e) **False**   These rise due to loss of feedback inhibition.

472 (a) **True**   Withdrawal causes menstruation.
   (b) **True**   Possibly due to structural affinity with mineralocorticoids.
   (c) **True**   By a negative feedback mechanism.
   (d) **True**   By suppressing pituitary gonadotrophins.
   (e) **False**   Oestrogens antagonize the effect of androgens on sebaceous glands and improve acne.

473 (a) **True**   Gonadal hormones are essential for their development.
   (b) **False**   Testosterone promotes secondary sexual characteristics even if the germ cells cannot develop.
   (c) **False**   They are delayed but appear eventually.
   (d) **False**   In most cases they develop.
   (e) **True**   These are required to stimulate the gonads to produce the essential sex hormones.

474 (a) **False**   This is the maximally fertile period.
   (b) **True**   This is difficult to reverse.
   (c) **True**   An interval is necessary before infertility can be assumed.
   (d) **True**   Intrauterine devices (IUDs) act in this way.
   (e) **False**   Oral contraceptives and intrauterine devices are more effective.

475 (a) **True**   Even though only one sperm ultimately fuses with the ovum.
   (b) **False**   Vasopressin and oxytocin are not needed for fertilization.
   (c) **False**   This would only reduce fertility moderately.
   (d) **True**   The higher temperature in the abdomen impairs function in the spermatogenic epithelium.
   (e) **False**   Male and female causes are about equally common in infertile couples.

476 (a) **True**   This is the threshold for postpartum haemorrhage.
   (b) **False**   After a long labour, the uterine contractions which normally close off the maternal sinusoids may be less efficient.
   (c) **True**   This also hinders sinusoidal compression.
   (d) **False**   Blood group O Rh-negative (universal donor) may be used.
   (e) **True**   This is a recognized complication (Sheehan's disease).

477 Failure to ovulate in a given cycle is likely if:
   (a) Pregnandiol appears in the urine in the second half of the cycle
   (b) Basal body temperature is constant throughout the cycle
   (c) Unilateral abdominal pain is experienced at mid-cycle
   (d) Cervical mucus showed evidence of unopposed oestrogen action in the second half of the cycle
   (e) The endometrium shows proliferating glands in the second half of the cycle

478 Features indicating poor physical condition in the newborn include:
   (a) A blue rather than a pale grey colour
   (b) A steadily rising heart rate
   (c) Spontaneous limb movements
   (d) Relaxed muscles with low tone
   (e) A grimace rather than a cough when the pharynx is stimulated

479 Premature labour is associated with a greater:
   (a) Risk of maternal complications
   (b) Risk of cerebral haemorrhage in the fetus
   (c) Fat content of the baby's skin
   (d) Fetal head to body size ratio
   (e) Need to feed the baby with milk

480 Following the menopause, the:
   (a) Vaginal secretions become more acid
   (b) Myometrium decreases in bulk
   (c) Libido may increase
   (d) Lack of sex hormones may produce general body changes
   (e) Level of pituitary gonadotrophins falls markedly

481 Women having their first child after the age of 35 have a greater:
   (a) Average blood loss than younger women
   (b) Incidence of ineffective uterine contractions during labour
   (c) Compliance of the perineum and vagina
   (d) Incidence of fetal abnormalities
   (e) Risk of spontaneous abortion

482 Infertility in the male can be explained by observations that:
   (a) There are no motile sperm in semen 15 min after ejaculation
   (b) 50% of the sperm in the semen are abnormal
   (c) The sperm count is 10^8/ml
   (d) The sperm count is 50% below average
   (e) There is widespread autonomic neuropathy

477 **(a)** **False** This is a normal derivative of progesterone which dominates the second half of the cycle.
    (b) **True** Normally there is a rise in the second half of the cycle.
    (c) **False** This is a sign of ovulation (mittelschmerz).
    (d) **True** Progesterone would normally prevent this effect ('ferning' due to salt crystals) in the second half of the cycle.
    (e) **False** Proliferating glands are normally present at this stage.

478 **(a)** **False** The pale grey colour suggests circulatory failure.
    (b) **False** This suggests recovery from vagal slowing.
    (c) **False** These are a good sign.
    (d) **True** This suggests severe hypoxia.
    (e) **True** This indicates depressed reflexes.

479 **(a)** **False** The fetus is smaller and is delivered more easily.
    (b) **True** Due to the greater fragility of cerebral veins at this stage of maturity.
    (c) **False** The skin tends to be brick red due to lack of fat.
    (d) **True** The body 'catches up' in late pregnancy.
    (e) **False** Milk is less well digested by premature infants.

480 **(a)** **False** They become less acid, infection is more likely.
    (b) **True** The uterus decreases in size.
    (c) **True** It may increase or decrease.
    (d) **True** Including osteoporosis and coronary artery disease.
    (e) **False** It rises due to loss of sex hormone negative feedback.

481 **(a)** **True** Uterine contraction is less effective in stopping bleeding.
    (b) **True** Again, due to deteriorating uterine function with age.
    (c) **False** It is less; Caesarean section is more often needed.
    (d) **True** Down's syndrome is one example of the genetic problems.
    (e) **True** Fetal abnormalities may be an important cause of this.

482 **(a)** **True** The sperm should be motile for at least 1 h.
    (b) **True** This indicates a serious defect in sperm formation.
    (c) **False** This is the normal value.
    (d) **False** The count must be below 10–20% to cause infertility.
    (e) **True** This neuropathy can affect the sexual reflexes.

483 Pregnant women with five or more previous deliveries have a greater risk of having:
   (a) Anaemia
   (b) An unfavourable presentation of the baby in the pelvis
   (c) Complications due to rhesus incompatibility
   (d) Serious loss of blood after delivery
   (e) Involuntary urination while coughing (stress incontinence)

484 Secretion of androgens in the adult female:
   (a) Is abnormal
   (b) In large amounts can cause enlargement of the clitoris
   (c) Does not affect the voice
   (d) May lead to growth of facial hair
   (e) May result in amenorrhoea

485 Fetal death is likely to result from serious impairment of fetal:
   (a) Liver function
   (b) Alimentary tract function such as obstruction
   (c) Renal function
   (d) Cerebral function
   (e) Cardiac function

483 **(a) True**   Due partly to depletion of iron stores.
    (b) True   Due to flabby uterine and abdominal wall muscles.
    (c) True   There have been more opportunities for sensitization.
    (d) True   Again due to deteriorating uterine function.
    (e) True   Due to pelvic damage.

484 **(a) False** Adrenal androgen secretion is normal.
    (b) True   It may grow to resemble a small penis.
    (c) False The voice deepens due to permanent laryngeal enlargement.
    (d) True   As in the male.
    (e) True   By suppressing the normal endometrial cycle.

485 **(a) False** The maternal liver can compensate.
    (b) False Fetal nutrition depends on the placenta.
    (c) False The maternal kidneys can compensate.
    (d) False Fetal survival does not depend on normal brain function.
    (e) True   Fetal circulation is needed to harness maternal support systems to supply fetal needs.

486 Ultrafiltration:
- (a) Separates the colloidal contents of a solution from the crystalloid contents
- (b) Is involved in the movement of water and electrolytes across cell membranes
- (c) Is involved in the formation of tissue fluid by the capillaries
- (d) Is involved in the formation of glomerular filtrate by the kidneys
- (e) Is involved in the formation of cerebrospinal fluid by the choroid plexuses

487 The Fick principle enables:
- (a) Blood flow through an organ to be calculated if organ uptake (U), arterial concentration (A) and venous concentration (V) are known for a given substance
- (b) Cardiac output to be calculated by injecting an indicator into the pulmonary artery and monitoring its concentration downstream in a systemic artery
- (c) Renal plasma flow to be calculated using PAH as the substance measured
- (d) Cardiac output to be estimated using the lungs as the organ and carbon dioxide as the substance measured
- (e) Cerebral blood flow to be measured using nitrous oxide as the substance taken up by the brain

488 Mitochondria:
- (a) Have membranes similar to the cell membrane
- (b) Are the chief site for protein synthesis
- (c) Are the chief site for generation of ATP
- (d) Are more numerous in brown than in white fat cells
- (e) Are absent near the membranes of actively secreting cells

489 Sleep is associated with:
- (a) An alpha rhythm in the electroencephalogram
- (b) Increased activity in the reticular activating system
- (c) A high level of vagal tone to the heart
- (d) Grinding movements of the teeth
- (e) A rise in central body temperature

490 During the Valsalva manoeuvre (forced expiration with glottis closed):
- (a) Pressure rises in the urinary bladder
- (b) The initial rise in arterial pressure coincides with a fall in the rate of venous return
- (c) Cardiac output rises
- (d) Heart rate slows
- (e) Peripheral resistance rises

486 (a) True   It depends on pore size and hydrostatic pressure gradients; colloids are larger particles than crystalloids.

   (b) False   Substances cross cellular walls by active transport or down electrochemical gradients; hydrostatic pressure gradients are not involved.

   (c) True   An outward hydrostatic pressure gradient exists at the arterial end of the capillary.

   (d) True   The pressure gradient in glomerular capillaries is greater than in other capillaries.

   (e) True   But active secretion is also involved in the formation of CSF.

487 (a) True   From the principle, $U = F(A - V)$, so $F = U/(A - V)$.

   (b) False   This method is related to the indicator dilution rather than the Fick principle.

   (c) True   Uptake = urine volume × urine concentration, V = zero; the formula is then the same as for PAH clearance.

   (d) True   Either oxygen or carbon dioxide can be the substance measured.

   (e) True   The Kety method to measure cerebral blood flow uses the Fick principle.

488 (a) True   Both membranes have the same lipid bilayer structure.

   (b) False   This applies to the ribosomes.

   (c) True   ATP is formed by oxidative phosphorylation.

   (d) True   Brown fat cells can generate energy, and hence heat, more rapidly.

   (e) False   They are concentrated where most energy is required.

489 (a) False   The alpha rhythm disappears during sleep.

   (b) False   Activity decreases; increased activity is associated with alertness.

   (c) True   This maintains a slow heart rate during sleep.

   (d) True   Teeth-grinding is associated with REM sleep and is called bruxism.

   (e) False   This tends to drop during the early hours of sleep.

490 (a) True   Due to the rise in intra-abdominal pressure.

   (b) True   Both result from the sudden initial rise in intrathoracic pressure.

   (c) False   It falls due to the reduced venous return.

   (d) False   It rises to compensate for a falling arterial pressure.

   (e) True   Another reflex compensatory mechanism to restore arterial pressure.

491 Jejunal mucosal cells are similar to proximal convoluted tubular cells in that both:
- (a) Absorb glucose by a process linked with sodium absorption
- (b) Absorb chloride ions actively
- (c) Absorb amino acids actively
- (d) Are rich in mitochondria
- (e) Possess microvilli on their luminal border

492 Exercise which doubles the metabolic rate is likely at least to double the:
- (a) Oxygen consumption
- (b) Cardiac output
- (c) Stroke volume
- (d) Arterial P_{CO_2}
- (e) Minute volume

493 The endoplasmic reticulum:
- (a) Is a complex system of intracellular tubules
- (b) Is a component of the Golgi apparatus
- (c) Has a membrane structure similar to the cell membrane
- (d) Is associated with ribonucleoprotein
- (e) Is well developed in secretory cells

494 In bone:
- (a) Osteoclasts are responsible for bone resorption
- (b) Osteoclasts are inhibited by parathyroid hormone
- (c) Maintenance of normal calcium content depends on exposure to mechanical stress
- (d) Strontium ions may substitute for calcium ions
- (e) The width of the epiphyseal plate is an index of its growth rate

495 The standard deviation of a series of observations
- (a) Is related to the scatter of the observations
- (b) Is such that about 95% of observations lie within one standard deviation of the mean
- (c) Should be calculated only if observations are normally distributed
- (d) Is a measure of the significance of the observations
- (e) Is a more valid expression of scatter than is the absolute range for small numbers of observations, e.g. less than five

496 Human circadian (24-h) rhythms:
- (a) Are triggered totally by external (exogenous) factors
- (b) Depend more on the integrity of the cerebral cortex than of the hypothalamus
- (c) Adapt within 48 h on changing from day to night shift work
- (d) For melatonin secretion produce high night-time and low day-time levels of the hormone
- (e) For the eosinophil count produce peak values around midday

491 **(a) True**   In both cases, glucose absorption is blocked by phlorhizin.
    (b) False   In both, these follow passively the absorption of sodium.
    (c) True   Again, the process requires active absorption of sodium.
    (d) True   Both expend considerable energy.
    (e) True   The two types of cell have very similar functions.

492 **(a) True**   Oxygen consumption is directly related to metabolic rate.
    (b) False   This rises more slowly than metabolic rate because tissue O_2 extraction increases.
    (c) False   Since heart rate rises, stroke volume does not rise in proportion to metabolic rate.
    (d) False   This is little changed.
    (e) True   This rises proportionately more than metabolic rate.

493 **(a) True**   It is analogous to the sarcoplasmic reticulum in muscle cells.
    (b) False   The Golgi apparatus differs in structure, function and location in the cell.
    (c) True   As do mitochondria.
    (d) True   RNA is found in the ribosomes which are attached to the cytoplasmic side of the tubules in rough endoplasmic reticulum and are responsible for protein synthesis.
    (e) True   It is the site of hormone and enzyme synthesis.

494 **(a) True**   They contain an acid phosphatase.
    (b) False   They are stimulated to resorb bone with release of calcium ions.
    (c) True   Calcium is released from bone during bed rest and inactivity.
    (d) True   This may subject bones to increased radioactivity.
    (e) True   The epiphyseal plate is relatively wide during childhood.

495 **(a) True**   It increases as scatter increases.
    (b) False   This is true for two standard deviations; for one standard deviation it is 66%.
    (c) True   Otherwise (b) above does not hold.
    (d) False   It is only a description of the scatter of observations.
    (e) False   It should not be calculated for very small numbers of observations.

496 **(a) False**   External triggers modulate an endogenous clock.
    (b) False   The hypothalamus is thought to be the location of the endogenous clock.
    (c) False   Adaptation takes about a week or longer.
    (d) True   Melatonin given at bedtime has been used to allay the circadian rhythm disturbances of sleep with jet-lag.
    (e) False   The peak is at night when cortisol level is minimal.

497 The mammalian cell membrane:
  (a) Is seen as an optically dense line using light microscopy
  (b) Consists mainly of protein
  (c) Is more permeable to fat-soluble than to water-soluble particles
  (d) Contains enzymes
  (e) Contains the receptors for steroid hormones

498 During early inspiration there is an increase in:
  (a) Heart rate
  (b) Central venous pressure
  (c) Intrapulmonary pressure
  (d) Abdominal girth
  (e) Afferent impulse traffic in the vagus nerves

499 Lysosomes:
  (a) Are membrane-bound organelles in the cytoplasm
  (b) Contain enzymes known as lysozyme
  (c) Are present in serous salivary glands
  (d) Enable neutrophil granulocytes to digest phagocytosed material
  (e) Can digest cellular contents

500 On lying down there is a decrease in the:
  (a) Central venous volume
  (b) Total systemic peripheral resistance
  (c) Ventilation:perfusion ratio in lung apices
  (d) Vital capacity
  (e) Rate of formation of urine

501 One mole of calcium ion
  (a) Is equivalent to 2 osmoles of calcium
  (b) Has the same mass as two equivalents of calcium ion
  (c) Is present in 1 litre of a normal solution of calcium ions
  (d) Weighs 40 g
  (e) Depresses the freezing point of water to the same extent as 1 mole of sodium ion

502 The Valsalva manoeuvre is followed by an decrease in:
  (a) Intrapleural pressure
  (b) Intra-abdominal pressure
  (c) Cardiac output
  (d) Arterial blood pressure
  (e) Heart rate

497 (a) **False** It is beyond the resolution of the light microscope.
  (b) **False** It consists mainly of lipid.
  (c) **True** Fat-soluble particles dissolve easily in the lipid matrix.
  (d) **True** These are involved in active transport.
  (e) **False** Steroid hormones cross the membrane and act intracellularly.

498 (a) **True** This is part of the cycle of respiratory sinus arrhythmia.
  (b) **False** It decreases due to expansion of the great veins.
  (c) **False** It falls due to expansion of the thoracic cage.
  (d) **True** Due to descent of the diaphragm.
  (e) **True** From stretch receptors in the lung mediating the Hering-Breuer reflex.

499 (a) **True** They are present in most cells.
  (b) **True** Lysozyme breaks down organic structures.
  (c) **True** Lysozyme in saliva helps to destroy ingested bacteria.
  (d) **True** The granules are examples of lysosomes.
  (e) **True** This is responsible for decomposition of tissues after death.

500 (a) **False** Volume expands as blood shifts from veins in the lower extremities to the chest.
  (b) **True** A reflex response to an increased central blood volume.
  (c) **True** Apical perfusion increases during recumbency.
  (d) **True** Due to vascular distension in the lungs and pushing up of the diaphragm.
  (e) **False** An increase in central blood volume results in increased urine formation.

501 (a) **False** One mole of any ion in solution exerts 1 osmole of osmotic pressure.
  (b) **True** Because calcium is divalent.
  (c) **False** A normal solution has one equivalent weight of ion/litre.
  (d) **True** This is the atomic weight in grams.
  (e) **True** One osmole of a substance in 1 litre of water depresses the freezing point by 1.86°C.

502 (a) **True** It returns to its normal negative value.
  (b) **True** It also returns to normal.
  (c) **False** It rises as blood surges back to the central circulation.
  (d) **False** It rises due to the great increase in cardiac output.
  (e) **True** A reflex response to the surge in arterial pressure.

503 Carbonic anhydrase has a role to play in the formation of:
- (a) HCl by the parietal cells of the stomach
- (b) Carbaminohaemoglobin
- (c) Cerebrospinal fluid in the choroid plexuses
- (d) Bicarbonate by the pancreas
- (e) Aqueous humour by the ciliary bodies

504 Sinus arrhythmia:
- (a) Refers to the cyclical changes in blood pressure that accompany the respiratory cycle
- (b) Can be observed in normal people
- (c) Has a greater amplitude in old than in young people
- (d) Is mediated mainly through sympathetic nerves to the heart
- (e) Can be used as an index of autonomic nerve function

505 Membrane ion channels:
- (a) Consist mainly of carbohydrate and lipid
- (b) Have a specific structure for each ion species
- (c) For sodium may be blocked by tetrodotoxin
- (d) May be opened by a given change in transmembrane potential
- (e) Remain open as long as the activating signal is present

506 The hydrogen ion concentration in a solution:
- (a) Equals the hydroxyl ion concentration at pH 7
- (b) Rises with its acidity
- (c) Of pH 5 is 100 times greater than that in a solution of pH 7
- (d) Is inversely related to the hydroxyl ion concentration
- (e) Is approximately doubled if the pH falls by 0.3 units

507 The human cell nucleus:
- (a) Has a membrane which is permeable to nucleic acid
- (b) In somatic cells contains 44 chromosomes
- (c) Stores its genetic material in the nucleolus
- (d) Has a skeleton of fine filaments
- (e) Is necessary for cell division

508 Athletic training leads to an increase in the:
- (a) Ratio of lean body mass to body fat
- (b) Resting vagal tone
- (c) Resting stroke volume
- (d) Maximal oxygen consumption
- (e) Blood lactate level for a given level of work

503 (a) **True** It facilitates H^+ and HCO_3^- formation from H_2O and CO_2, and H^+ is secreted.

    (b) **False** CO_2 combines directly with amino groups to form carbaminohaemoglobin.

    (c) **False** Blockade of the enzyme does not affect CSF formation.

    (d) **True** It facilitates H^+ and HCO_3^- formation from CO_2 and H_2O, and HCO_3^- is secreted.

    (e) **True** Blockade of the enzyme reduces the rate of formation of aqueous humour.

504 (a) **False** It refers to the cyclical changes in heart rate that accompany the respiratory cycle.

    (b) **True** It is a normal, harmless phenomenon.

    (c) **False** It decreases in amplitude with age.

    (d) **False** It is abolished by vagal blockade by atropine

    (e) **True** It is impaired in vagal autonomic neuropathy.

505 (a) **False** They are usually proteins.

    (b) **False** There is usually a family of channels for each ion species.

    (c) **True** This toxin from the puffer fish specifically blocks sodium channels.

    (d) **True** These channels are 'voltage-gated' channels; 'ligand-gated' channels are opened by chemical substances such as acetylcholine.

    (e) **False** They close automatically after a certain interval.

506 (a) **True** This is the neutral pH of water.

    (b) **True** The pH falls

    (c) **True** pH is minus log_{10} hydrogen ion concentration.

    (d) **True** The product of these ionic concentrations is constant.

    (e) **True** Log 2 = 0.301.

507 (a) **True** Messenger RNA crosses the nuclear membrane.

    (b) **False** The total is 46; i.e. 44 + XX or XY.

    (c) **False** The nucleolus is a condensation of RNA; the genetic material is in chromosomes.

    (d) **True** These are attached to the nuclear membrane.

    (e) **True** For example, red blood cells cannot divide.

508 (a) **True** Skeletal muscle bulk increases with activity more than fat.

    (b) **True** This explains the slow pulse rate seen in athletes.

    (c) **True** This maintains resting cardiac output with the slower pulse rate.

    (d) **True** This is a good index of physical fitness.

    (e) **False** Due to greater capacity to deliver oxygen to muscle, the reverse is true.

509 The total osmotic pressure of human plasma is:
  (a) About 25 mmHg
  (b) Similar to that of 0.9% saline
  (c) Similar to that of 0.9% glucose solution
  (d) Opposing the tendency of fluid to leave capillaries
  (e) Equal to that of intracellular fluid

510 With increasing age there is a fall in the:
  (a) Arterial pulse pressure
  (b) Alveolar–arterial pressure gradient for oxygen
  (c) Ability of the kidneys to concentrate urine
  (d) Residual volume of the lungs
  (e) Fasting blood glucose concentration

511 Ingestion of protein:
  (a) Raises metabolic rate more than ingestion of equal calorific amounts of fat or carbohydrate
  (b) Tends to lower the pH of urine
  (c) Permits the body to synthesize the essential amino acids
  (d) Yields more toxic metabolites than fat or carbohydrate
  (e) Should exceed 2 g/kg body weight/day to satisfy normal body requirements

512 Drinking 1 litre of water:
  (a) Increases secretion of antidiuretic hormone
  (b) Reduces the plasma sodium concentration
  (c) Causes more osmolar change in portal venous than in systemic venous blood
  (d) Causes body cells to shrink
  (e) Decreases the specific gravity of the body

513 The Golgi apparatus is:
  (a) Found in all eukaryotic cells
  (b) A collection of complex tubules and vesicles
  (c) Well developed in cells with secretory activity
  (d) Associated with endoplasmic reticulum
  (e) Not conspicuous in neurones

514 An acid–base buffer system:
  (a) Can be a mixture of a weak acid and its conjugate base
  (b) Can be a solution of sodium and bicarbonate ions
  (c) Prevents any change in pH when acid is added
  (d) Works best when acid and base are equal in concentration
  (e) With pK = 4 would be a better blood buffer than one with pK = 6

509 (a) **False** It is about 5500 mmHg (25 mmHg is the colloid osmotic pressure).
   (b) **True** Both have an osmolality around 290 mosmol/litre.
   (c) **False** It is similar to 5% glucose which contains about the same number of particles.
   (d) **False** The colloid osmotic pressure does this; the crystalloids pass capillary walls freely and exert no effective pressure.
   (e) **True** Since water can cross the cell membrane freely.

510 (a) **False** It rises with loss of arterial elasticity.
   (b) **False** It rises as more of the lung is underventilated.
   (c) **True** Due to a fall in maximal renal medullary osmolality.
   (d) **False** It increases as small airways collapse more readily.
   (e) **False** It rises as glucose homeostasis loses efficiency.

511 (a) **True** Due, perhaps, to the additional metabolic work in processing protein in the body.
   (b) **True** Protein is the main dietary source of acidic residues excreted by the kidney.
   (c) **False** Essential amino acids cannot be synthesized in the body.
   (d) **True** These metabolites, normally detoxified in the liver, may cause hepatic encephalopathy in hepatic failure.
   (e) **False** 1 g/kg is adequate.

512 (a) **False** It suppresses secretion of antidiuretic hormone.
   (b) **True** Osmolality is reduced in all the body fluid compartments.
   (c) **True** The water is absorbed via the portal vein.
   (d) **False** They swell as water is drawn in osmotically.
   (e) **False** It moves it upwards from its value of less than 1.0.

513 (a) **True** Usually close to the nucleus.
   (b) **False** It is a collection of about six flattened, membrane-enclosed sacs stacked together.
   (c) **True** Secretions are packaged into vesicles in the Golgi apparatus.
   (d) **True** Proteins formed by the granular endoplasmic reticulum fuse with the membranes of the Golgi sacs before being packed into vesicles and released into the cytoplasm.
   (e) **False** Neurones are secretory cells.

514 (a) **True** Or a mixture of a weak base with its conjugate acid.
   (b) **False** It can be a mixture of carbonic acid and bicarbonate ions.
   (c) **False** It minimizes, but does not prevent, pH change.
   (d) **True** pH = pK + log [base]/[acid]; log 1 = 0.
   (e) **False** pK should be near blood pH, which is 7 4.

515 Rapid eye movement (REM) sleep is:
   (a) Associated with EEG waves of high amplitude
   (b) Less common after a period of sleep deprivation
   (c) Associated with a high level of general muscle tone
   (d) Ineffective in relieving fatigue
   (e) A more frequent component of sleep in the elderly than in the young

516 From childhood to old age:
   (a) There is a steady decrease in total sleeping time per day
   (b) Deep (stage 4) sleep increases as a percentage of total daily sleep
   (c) Brain mass steadily decreases
   (d) Body water decreases as a percentage of body mass
   (e) Sleep becomes less aggregated into a single sleeping period

517 Immersion of an upright adult to chest level in water increases the:
   (a) Water pressure on the feet to nearer 100 than 20 mmHg
   (b) Thoracic blood volume
   (c) Total peripheral resistance
   (d) Sodium excretion in the urine
   (e) Transmural pressure in the blood vessels of the feet

518 In the days following a major surgical operation there is:
   (a) An increase in plasma cortisol level
   (b) A negative nitrogen balance
   (c) Potassium retention
   (d) A negative sodium balance
   (e) A decreased tendency for the blood to clot

519 The graph of body mass (ordinate) versus age:
   (a) Normally steepens from birth to 5 years
   (b) Normally steepens during secondary sexual development
   (c) Steepens earlier than usual with precocious puberty
   (d) Is parallel for males and females
   (e) Deviates towards the normal curve with successful treatment of a child
      with dwarfism

520 Obesity can be treated successfully by:
   (a) Diuretic drugs
   (b) Supplementing the normal diet with slimming foods
   (c) Confining the diet to foods of low calorific value
   (d) Increasing exercise without increasing food intake
   (e) Reducing food intake relative to energy expenditure

521 An injection of atropine typically produces a decrease in:
   (a) Resting heart rate
   (b) Skeletal muscle strength
   (c) Salivary flow
   (d) Mucus secretion in the airways
   (e) Impulse transmission at autonomic ganglia

515 (a) **False** The waves are of high frequency and low amplitude.
   (b) **False** REM sleep increases proportionately after sleep deprivation.
   (c) **False** General muscle tone is reduced.
   (d) **False** Fatigue is poorly relieved if REM sleep is prevented.
   (e) **False** It occurs less frequently in the elderly.

516 (a) **True** The elderly seem to need less sleep.
   (b) **False** This percentage also falls steadily.
   (c) **False** It increases markedly in early childhood.
   (d) **True** From over 80% in the fetus to less than 50% in old age.
   (e) **True** Though aggregation increases in infancy, it later declines steadily.

517 (a) **True** 100 mmHg is equivalent to 1.36 m of H_2O.
   (b) **True** Due to pressure on capacity vessels in the lower body.
   (c) **False** Increased central blood volume reflexly lowers resistance.
   (d) **True** Increased central blood volume causes diuresis, perhaps via atrial natriuretic hormone.
   (e) **False** It decreases transmural pressure and so protects the vessels against distension due to gravitational forces.

518 (a) **True** This is essential for normal recovery.
   (b) **True** Cortisol breaks down protein to form glucose.
   (c) **False** Potassium excretion increases.
   (d) **False** There is a positive sodium balance.
   (e) **False** The clotting tendency increases for about 10 days.

519 (a) **False** It becomes less steep over this period.
   (b) **True** This is the adolescent growth spurt.
   (c) **True** Sex hormones produce an early adolescent growth spurt.
   (d) **False** Females show an earlier adolescent growth spurt.
   (e) **True** The faster than normal growth rate is an important sign of appropriate treatment.

520 (a) **False** This reduces body water but not body fat.
   (b) **False** This makes the person fatter.
   (c) **False** This is ineffective unless the total energy intake is reduced.
   (d) **True** Energy expenditure increases; energy intake is unchanged.
   (e) **True** This is the only effective long-term treatment.

521 (a) **False** It increases due to blocking of acetylcholine's action at the sinus node.
   (b) **False** Atropine does not block acetylcholine's action on skeletal muscle.
   (c) **True** Salivation is stimulated by cholinergic autonomic nerves.
   (d) **True** This may be useful during surgical operations.
   (e) **False** This cholinergic transmission is resistant to atropine.

522 In percentage terms, arterial P_{O_2} is more affected than arterial O_2 content by:
    (a) Carbon monoxide poisoning
    (b) Anaemia
    (c) A 20% fall in inspired P_{O_2}
    (d) Ascent to 2000 m (about 6500 feet) above sea level
    (e) Breathing oxygen at three atmospheres pressure

523 Polycythaemia is caused by:
    (a) Lung disease which causes a fall in oxygen saturation in arterial blood
    (b) Heart disease which causes a right to left shunt
    (c) Chronic renal failure
    (d) Pregnancy
    (e) High doses of vitamin B_{12}

524 When kept afloat by a life jacket, survival time in water at 15°C is:
    (a) Usually 12–24 h
    (b) Limited by muscular fatigue
    (c) Extended if many layers of clothing are worn
    (d) Extended by swimming gently rather than floating motionless
    (e) More prolonged in fat than in thin persons

525 An organ transplant is less likely to be rejected if the recipient:
    (a) Is given glucocorticoid treatment
    (b) Has previously received a skin graft from the same individual
    (c) Is an infant
    (d) Is an identical rather than a non-identical twin of the donor
    (e) Has the same blood group as the donor

526 Obesity:
    (a) Is associated with reduced life expectancy
    (b) Is one cause of diabetes mellitus
    (c) In parents is associated with obesity in their children
    (d) Is usually due to an endocrine disorder
    (e) May be treated by surgical removal, or bypass, of part of the small intestine

527 Sudden exposure to an atmospheric pressure of 100 mmHg (13 kPa), as might occur with loss of aircraft cabin pressure at around 15 000 m altitude, causes:
    (a) Rupture of the ear drums by outward bulging
    (b) The appearance of gas bubbles in joints and lungs
    (c) No serious fall in the alveolar oxygen pressure if the person is breathing pure oxygen
    (d) Expansion of gas in closed spaces in the body
    (e) Gradual loss of consciousness in 10–15 min due to hypoxia

522 (a) **False** Pressure is little affected but content is decreased.
  (b) **False** Pressure is normal but content decreased.
  (c) **True** Arterial P_{O_2} is reduced but content falls by only 3% or so because of the shape of the oxygen dissociation curve.
  (d) **True** This causes a fall of about 20% in inspired oxygen pressure.
  (e) **True** This causes a huge rise in P_{O_2} but a relatively small increase in arterial oxygen content since the haemoglobin is fully saturated with oxygen.

523 (a) **True** This stimulates production of erythropoietin.
  (b) **True** This also reduces the oxygen content of arterial blood.
  (c) **False** The erythropoietin-producing cells are damaged.
  (d) **False** The increase in plasma volume causes a fall in red cell count.
  (e) **False** This does not raise red cell concentration above normal.

524 (a) **False** It is usually under 2 h.
  (b) **False** It is limited by hypothermia.
  (c) **True** This improves insulation by trapping layers of water around the body to form a stable microclimate.
  (d) **False** Swimming increases heat loss by disturbing the microclimate.
  (e) **True** Due to better thermal insulation by the superficial fat.

525 (a) **True** This suppresses immune reactions.
  (b) **False** This would sensitize the immune response.
  (c) **True** The immune system is less well developed in infants.
  (d) **True** Only identical twins have identical genes and hence identical tissue antigens.
  (e) **True** However, blood group compatibility is not a reliable index of tissue compatibility.

526 (a) **True** Being 20% overweight reduces life expectancy by about 20%.
  (b) **True** The diabetes may then be cured by weight reduction.
  (c) **True** Both genetic and environmental factors may operate.
  (d) **False** This is a very rare cause.
  (e) **True** This leads to malabsorption, but the resulting malnutrition may have side-effects.

527 (a) **True** Ambient pressure drops much faster than middle ear pressure.
  (b) **True** This is a manifestation of decompression sickness.
  (c) **False** More than half of the total alveolar pressure will be taken up by carbon dioxide and water vapour.
  (d) **True** This must happen from Boyle's law.
  (e) **False** Sudden severe hypoxia of this order causes unconsciousness in less than a minute.

528 Restoration of the blood volume after haemorrhage is aided by:
  (a) Contraction of venous reservoirs
  (b) A fall in capillary pressure in certain vascular beds
  (c) Arteriolar vasoconstriction
  (d) Mobilization of intracellular fluid into the circulation
  (e) An increase in the osmotic pressure of the plasma proteins

529 A transplanted kidney is:
  (a) Able to maintain the recipient's blood urea at normal levels
  (b) Able to correct the anaemia of chronic renal failure
  (c) Probably being rejected if the glomerular filtration rate is falling rapidly
  (d) Probably being rejected if there is a sharp rise in body temperature in the absence of infection
  (e) Probably being rejected if the urinary volume is low and the osmolality high

530 Helium is used to replace nitrogen in gas breathed by divers because it:
  (a) Is more soluble in body fluids
  (b) Diffuses through the tissues more rapidly
  (c) Causes less depression of cerebral function
  (d) Diminishes the work of breathing relative to nitrogen
  (e) Combines less readily with haemoglobin

531 Someone who has received an electric shock causing ventricular fibrillation:
  (a) Loses consciousness in less than a minute
  (b) Has a reduction in cardiac output of about 50%
  (c) Has a rapid but weak carotid pulse
  (d) Should be given external cardiac massage after removal from the electrical contact
  (e) Should not be given artificial ventilation until the ventricular fibrillation has been reversed

532 A patient with fever:
  (a) Has warm extremities as central temperature rises
  (b) Has a raised basal metabolic rate
  (c) Shows evidence of altered hypothalamic function
  (d) Loses the capacity for reflex thermoregulation
  (e) May develop heat stroke if the core temperature rises to 40°C

533 Acidosis in a patient may lead to:
  (a) Increased urinary excretion of potassium
  (b) Hypoventilation
  (c) A blood pH of less than 5.5
  (d) A urinary pH of less than 5.5
  (e) Tetany

528 (a) False This redistributes but does not increase blood volume.
   (b) True This favours absorption of tissue fluid into the circulation.
   (c) True This leads to the fall in downstream capillary pressure.
   (d) False Haemorrhage does not increase the osmolality of extracellular fluid.
   (e) False Colloid osmotic pressure falls as tissue fluid is drawn into the circulation.

529 (a) True One normal kidney can do this.
   (b) True By replacing the deficient erythropoietin.
   (c) True This suggests nephron damage.
   (d) True This, like a raised ESR and granulocyte count, suggests tissue destruction.
   (e) False A high urinary osmolality suggests excellent renal function.

530 (a) False Being less soluble, less goes into solution during compression so there is less bubble formation during decompression.
   (b) True This also reduces the time needed for decompression.
   (c) True It is less narcotic than nitrogen.
   (d) True It is less viscous than nitrogen.
   (e) False Neither combines with haemoglobin.

531 (a) True Due to the abrupt fall in cerebral blood flow.
   (b) False Cardiac output is nil.
   (c) False There is no cardiac output; hence no pulse.
   (d) True This will prevent brain damage until the fibrillation has been reversed.
   (e) False Artificial ventilation must accompany cardiac massage to avoid brain damage.

532 (a) False Reflex vasoconstriction causes cold hands at this stage.
   (b) True The raised body temperature speeds metabolism.
   (c) True The 'set point' for temperature regulation is raised.
   (d) False Core temperature is maintained around the raised level.
   (e) False The core temperature needs to rise to about 43°C before heat stroke develops.

533 (a) False Hydrogen ions compete with potassium for secretion.
   (b) False Ventilation is increased in acidosis.
   (c) False This level would be fatal.
   (d) True Urinary pH may fall below 5.0
   (e) False Acidosis reduces the risk of tetany by decreasing protein affinity for calcium.

534 When a patient inherits a disease as a recessive autosomal character:
  (a) One of the parents of the patient will exhibit the disease
  (b) All of the children of the patient will exhibit the disease
  (c) All of the children of the patient will be carriers
  (d) Both parents of the patient must carry the recessive character
  (e) Subsequent siblings have a 50% risk of the disorder

535 If treatment A is superior to treatment B in certain respects (with a P value less than 0.01) it can be concluded that:
  (a) A given patient's chances of improvement by treatment A are 99%
  (b) The likelihood that the observed difference between treatments A and B is due to chance is less than 1%
  (c) The observed superiority is statistically significant
  (d) At least 100 patients were studied
  (e) Treatment A should now be substituted for treatment B

536 A low serum potassium level:
  (a) Can be suspected from the appearance of the ECG
  (b) Can result from repeated vomiting of gastric contents
  (c) Indicates that total body potassium is low
  (d) May be a consequence of aldosterone deficiency
  (e) Impairs gut motility

537 A rise in the osmolality of extracellular fluid may lead to:
  (a) Thirst
  (b) Increased water reabsorption in the proximal convoluted tubules
  (c) Release of vasopressin
  (d) A fall in intracellular fluid volume
  (e) Suppression of sweat secretion

538 Maximal exercise leads to an increase in:
  (a) Systolic blood pressure
  (b) Total peripheral resistance
  (c) Blood lactic acid
  (d) Tissue fluid formation in active muscles
  (e) Urinary output

539 Following adaptation to a hot climate there is an increase in:
  (a) Basal metabolic rate
  (b) Resting cardiac output
  (c) Urinary output
  (d) The ability to lose heat by sweating
  (e) Exercise tolerance

534 (a) **False** Both parents are usually healthy.
    (b) **False** They are likely to be normal due to the dominant normal gene contributed by the spouse.
    (c) **True** All will receive the recessive gene from the patient.
    (d) **True** Only thus will the patient have a homozygous genotype.
    (e) **False** From simple Mendelian laws the chances are 25%.

535 (a) **False** This cannot be concluded from the results given.
    (b) **True** This is the meaning of P less than 0.01.
    (c) **True** Conventionally P must then be less than 0.05.
    (d) **False** This cannot be concluded from the results given.
    (e) **False** Adverse effects may outweigh the benefit shown.

536 (a) **True** It leads to low-voltage T waves.
    (b) **True** The vomitus is relatively rich in potassium.
    (c) **False** It is a poor reflector of total body potassium.
    (d) **False** This favours retention of potassium and a high serum level.
    (e) **True** It may cause paralytic ileus (paralysis of peristalsis).

537 (a) **True** Due to stimulation of hypothalamic osmoreceptors.
    (b) **False** Water reabsorption in proximal tubules is not geared to meet body water needs.
    (c) **True** Impulses from hypothalamic osmoreceptors lead to release of vasopressin from the posterior pituitary gland.
    (d) **True** The cells shrink as fluid is drawn out osmotically.
    (e) **False** Sweat production is not geared to body fluid requirements.

538 (a) **True** Due mainly to the increase in the force of ventricular ejection.
    (b) **False** This falls markedly due to vasodilation in muscle, heart and skin.
    (c) **True** Since the muscles may have to work in anaerobic conditions.
    (d) **True** This results from the rise in hydrostatic pressure in muscle capillaries.
    (e) **False** This decreases, probably due to reflex vasoconstriction in the kidneys and increased release of vasopressin.

539 (a) **False** BMR falls due to a decrease in resting thyroid activity.
    (b) **True** Partly due to increased thermoregulatory blood flow.
    (c) **False** Urinary output is usually reduced due to increased water loss by extrarenal routes.
    (d) **True** Daily secretion of sweat may rise to several litres.
    (e) **True** Due mainly to the improved ability to dissipate heat.

540 Sudden complete obstruction of the respiratory tract causes:
   (a) A fall in blood pressure
   (b) Stimulation of central chemoreceptors
   (c) Cyanosis
   (d) Reflex apnoea
   (e) Dilation of the pupils

541 Inherited diseases associated with sex-linked recessive genetic disorders:
   (a) Involve the Y rather than the X chromosome
   (b) Are more common in females than in males
   (c) Are transmitted by the female but not by the male
   (d) May fail to manifest themselves in female carriers
   (e) Include haemophilia

542 Infants differ from adults in that their:
   (a) Nitrogen balance is normally positive
   (b) Extracellular fluid volume is a larger proportion of total body water
   (c) Blood contains reticulocytes
   (d) Total peripheral resistance is lower
   (e) Brown fat stores are relatively small

543 Excessive sweating (hyperhidrosis) in a limb is likely to be relieved by blockade
   of impulse transmission in:
   (a) Autonomic ganglia with ganglion-blocking drugs
   (b) Autonomic cholinergic nerve endings with atropine
   (c) Somatic cholinergic nerve endings with curare
   (d) Adrenergic nerve endings with phentolomine
   (e) Sympathetic motor nerves supplying the limb by surgical sympathectomy

544 A disease inherited as a dominant autosomal character:
   (a) Affects males and females equally
   (b) Affects all the children of the affected adult
   (c) Usually prevents reproduction
   (d) Requires that both parents carry the abnormality
   (e) May be transmitted by a carrier who does not manifest the disease

545 The effects of moving from sea level to an altitude of 5000 m include an
   increase in:
   (a) Alveolar ventilation
   (b) Blood bicarbonate level
   (c) Appetite for food
   (d) Exercise tolerance
   (e) Simulation of the bone marrow

540 **(a)** **False**   Blood pressure rises due to reflex stimulation of the heart and peripheral vasoconstriction.

(b) **True**   Accumulation of CO_2 is mainly responsible for the reflex cardiovascular and respiratory responses.

(c) **True**   Deoxygenated haemoglobin appears in the arterial blood.

(d) **False**   Respiratory effort increases with chemoreceptor stimulation.

(e) **True**   Part of the generalized sympathetic response to stress.

541 **(a)** **False**   The abnormality is on the X chromosome.

(b) **False**   Females are protected by the normal X chromosome.

(c) **False**   Both can transmit a defective X chromosome.

(d) **True**   Due to protection by a normal X chromosome.

(e) **True**   Colour blindness is another example.

542 **(a)** **True**   Adults are normally in nitrogen balance.

(b) **True**   It exceeds intracellular fluid volume.

(c) **False**   Both infants and adults have reticulocytes in the blood.

(d) **False**   It is higher; the pressure gradient to cardiac output ratio is much greater.

(e) **False**   They are relatively much larger.

543 **(a)** **True**   But the side-effects are likely to be worse than the disease.

(b) **True**   Again side-effects may be troublesome.

(c) **False**   This paralyses skeletal muscles without inhibiting sweating.

(d) **False**   The nerves responsible for sweating are cholinergic.

(e) **True**   This is an effective therapy.

544 **(a)** **True**   The autosomes are similar for males and females.

(b) **False**   Half would receive the normal autosome.

(c) **False**   Such genes are commonly transmitted.

(d) **False**   If so, it would not be a dominant disorder.

(e) **False**   Carriers of a dominant character exhibit the disease.

545 **(a)** **True**   Due to stimulation of peripheral chemoreceptors by hypoxia.

(b) **False**   Bicarbonate is lost in the urine to compensate for the respiratory alkalosis.

(c) **False**   Loss of appetite (anorexia) is a common complaint in mountain climbers.

(d) **False**   Exercise tolerance is reduced by the decreased ability to deliver O_2 to the blood.

(e) **True**   Due to an increased erythropoietin level.

546 The ratio of intravascular hydrostatic pressure to colloid osmotic pressure is greater:
  (a) In splanchnic than in renal glomerular capillaries
  (b) Than normal in patients with hepatic failure
  (c) Than normal in the systemic capillaries of patients following severe blood loss
  (d) Than normal in capillaries where there is oedema due to venous obstruction
  (e) In systemic than in pulmonary capillaries

547 Sudden application of cold water to the:
  (a) Hand causes local vasoconstriction
  (b) Hand causes vasodilation in the opposite hand
  (c) Oesophagus can cause changes in the electrocardiogram
  (d) External auditory meatus causes nausea and nystagmus
  (e) Whole body by immersion causes apnoea

548 Normal healthy young adults can tolerate loss of half of their:
  (a) Renal tissue without developing renal failure
  (b) Pulmonary tissue without developing respiratory failure
  (c) Circulating platelets without developing a haemorrhagic tendency
  (d) Seminal sperm count without suffering from infertility
  (e) Liver without developing hepatic failure

546 (a) False It is higher in glomerular capillaries, as is necessary for filtration.
  (b) True Colloid osmotic pressure falls because the liver fails to synthesize enough albumin.
  (c) False The reverse is true, causing transfer of tissue fluid into the circulation.
  (d) True Hence excess fluid is forced out of the capillaries into the interstitial spaces.
  (e) True The lower ratio in the lungs normally prevents fluid leak into the alveoli.

547 (a) True Due to the direct effect of cold on vascular smooth muscle.
  (b) False Immersion of a hand in cold water provokes general vasoconstriction and a rise in blood pressure – the cold pressor response.
  (c) True By cooling the myocardium.
  (d) True By inducing currents in endolymph which stimulate vestibular receptors.
  (e) False By stimulating uncontrollable gasping under water, it may result in drowning

548 (a) True Normal function can be maintained with one kidney.
  (b) True Though maximum exercise tolerance is reduced, considerable exertion is possible.
  (c) True The platelet count must fall by more than 75% before bleeding problems arise.
  (d) True Infertility is unlikely unless the count falls to around a quarter of the normal value.
  (e) True In short, most body functions carry at least a 50% reserve in the young adult.

549 In athletes, physical fitness is more closely correlated with:
  (a) Maximal oxygen uptake than with resting oxygen uptake
  (b) Maximal pulse rate than with resting pulse rate
  (c) Maximal minute ventilation than with maximal cardiac output
  (d) Blood oxygen saturation than with blood lactate level during strenuous exercise
  (e) Resting vagal tone than with resting sympathetic tone to the heart

550 The muscle fibres adapted to endurance running:
  (a) Are classified as slow rather than fast
  (b) Have a relatively high myoglobin content
  (c) Are red rather than white
  (d) Have a relatively high mitochondria content
  (e) Are classified as anaerobic rather than aerobic

551 The oxygen consumed per minute:
  (a) Is greater than the carbon dioxide produced per minute during long-distance running
  (b) In the resting adult is nearer 100 than 150 ml
  (c) During intense mental activity can rise to twice the resting level
  (d) During brisk walking is nearer five times than twice the resting level
  (e) In an Olympic athlete can rise to 50 litres

552 The increase in blood flow to muscle in an exercising limb is related to a rise in:
  (a) Local P_{CO_2}
  (b) Local H^+ concentration
  (c) Local muscle temperature
  (d) Arterial pressure
  (e) Vasodilator nerve activity

553 During muscular training:
  (a) Neural control factors improve performance before there is evidence of skeletal muscle hypertrophy
  (b) Repeated stretching of skeletal muscle fibres leads to their hypertrophy
  (c) There is a gradual decrease in the size of the heart in diastole
  (d) There is a gradual increase in the O_2 extraction rate from blood perfusing exercising skeletal muscle
  (e) The increase in skeletal muscle blood flow for a given work load decreases

554 Blood lactic acid is:
  (a) Normally undetectable in resting subjects
  (b) A product of anaerobic metabolism
  (c) Increased by a 100 m dash
  (d) Not increased during steady-state running in a marathon race
  (e) Raised to about 5–10 mol/litre during maximal exercise

549 (a) **True**   Maximal oxygen uptake is the 'gold standard' of fitness.
     (b) **False**   Maximal pulse rate is related to age; a slow resting pulse indicates fitness.
     (c) **False**   Cardiac output is the usual limiting factor in exercise.
     (d) **False**   Saturation changes little during exercise; a relatively low lactate level during strenuous exercise is an indication of fitness
     (e) **True**   A high resting vagal tone is the cause of the low resting heart rate.

550 (a) **True**   The time course of their twitch is relatively long.
     (b) **True**   This provides an intracellular store of oxygen.
     (c) **True**   This is due to their myoglobin content.
     (d) **True**   They generate energy by oxidative phosphorylation.
     (e) **False**   Their high rate of oxygen consumption classifies them as aerobic.

551 (a) **True**   The respiratory quotient is less than 1.0 during long-distance running.
     (b) **False**   It is around 250–300 ml.
     (c) **False**   Brain oxygen consumption can be redistributed but the total changes little.
     (d) **True**   This is a useful way of maintaining fitness, particularly in older people.
     (e) **False**   The maximum recorded is under 10 litres.

552 (a) **True**   This causes local vasodilation.
     (b) **True**   Acidosis also favours vasodilation.
     (c) **True**   Heat dilates blood vessels.
     (d) **False**   Exercising a limb causes little change in mean arterial pressure.
     (e) **False**   Vasomotor nerves are not involved in exercise hyperaemia.

553 (a) **True**   Neural factors account for most of the early improvement in performance.
     (b) **True**   This is a potent factor causing hypertrophy.
     (c) **False**   There is an increase due to heart muscle hypertrophy and increased diastolic filling with the lower resting heart rate.
     (d) **True**   This results in a lower blood lactate level for a given work load.
     (e) **True**   A consequence of the greater oxygen extraction rate.

554 (a) **False**   The resting level is around 1 mmol/litre.
     (b) **True**   It is a marker for anaerobic metabolism.
     (c) **True**   Almost all the energy used in the 100 m dash comes from anaerobic metabolism.
     (d) **False**   Even though the lactic acid level is steady, it is still raised during steady-state exercise.
     (e) **False**   It rises to 5–10 mmol/litre.

555 Isotonic (dynamic) exercise differs from isometric (static) exercise in that there is less:
  (a) Increase in systolic arterial pressure
  (b) Increase in diastolic arterial pressure
  (c) Assistance to the circulation by the muscle pump
  (d) Use of slow-twitch muscle fibres
  (e) Reliance on anaerobic glycolysis

556 Electrocardiological danger signs during incremental treadmill exercise include:
  (a) A heart rate equal to the maximal predicted for the person's age
  (b) An R-R interval of about 500 ms
  (c) R waves with an amplitude greater than 1 mV
  (d) Ventricular tachycardia
  (e) ST depression of 1 mm

557 Exercising in a hot chamber may induce:
  (a) Fainting due to a decreased total peripheral resistance
  (b) Heat stroke when core temperature rises above 40°C
  (c) A rise in alveolar P_{CO_2}
  (d) A decrease in the osmolality of extracellular fluid
  (e) Heat adaptation if performed daily for several weeks

558 Cold:
  (a) Injury to feet exposed for long periods to 5–10°C is due to frost bite
  (b) Injury to the extremities is made less likely by increased affinity of haemoglobin for O_2 at low temperatures
  (c) Environments may induce a five-fold rise in resting metabolic rate
  (d) Water immersion causes death from hypothermia more rapidly in fat than in thin people
  (e) Water immersion of the hand at 5°C is painless

559 The maximum possible metabolic rate during exercise is:
  (a) Reached when the blood lactate level starts to fall
  (b) Reached when the respiratory exchange ratio starts to fall
  (c) Reached when ventilation reaches the maximum breathing capacity
  (d) Reduced by about half if the haemoglobin level falls by half
  (e) About 50 times the resting rate in an athlete

560 Asthma can interfere with exercise by:
  (a) Increasing the work of breathing
  (b) Cold-induced bronchial muscle spasm
  (c) Limiting alveolar ventilation
  (d) Reducing the diffusing capacity of the lung alveoli
  (e) Reducing the oxygen-carrying capacity of the blood

555 (a) False There is a greater increase during isotonic exercise.
  (b) True Diastolic pressure changes little with isotonic exercise but rises with isometric exercise.
  (c) False The muscle pump does not function during static exercise.
  (d) False In dynamic exercise more use is made of endurance fibres.
  (e) True Isometric exercise relies mainly on anaerobic glycolysis.

556 (a) False It is normal to reach one's predicted maximal heart rate.
  (b) False This implies a heart rate of about 120 beats/min, well within the normal range.
  (c) False This is also a normal finding.
  (d) True Stroke volume is impaired and ventricular fibrillation may develop.
  (e) False Depression of 2–3 mm is the borderline level for danger.

557 (a) True Vasodilation in skin, in addition to that in muscle, may reduce arterial pressure to fainting point.
  (b) True There is a serious risk of a progressive rise to fatal levels due to positive feedback.
  (c) False It falls due to the hyperventilation induced by the rise in core temperature.
  (d) False It rises due to the high output of hypotonic sweat.
  (e) True By improving the efficiency of heat-losing mechanisms, heat adaptation allows the subject to have a smaller rise in core temperature for a given work load.

558 (a) False The tissues do not freeze at this temperature; 'trench foot' injury can occur.
  (b) False Hypoxia is an increased risk due to poor release of oxygen to the tissues.
  (c) True Increased muscle tone and shivering account for this.
  (d) False Fat people have much better insulation of their body core.
  (e) False It is very painful, a warning of the danger of such temperatures.

559 (a) False The lactate level keeps rising with increasing severity of exercise.
  (b) False Increasing lactate acid increases ventilatory elimination of CO_2; a rising ratio suggests that exercise is approaching maximal.
  (c) False The level of ventilation in exercise does not reach the maximum breathing capacity.
  (d) True Oxygen delivery depends crucially on the haemoglobin level.
  (e) False The limit is about half of this.

560 (a) True Due to the increased resistance of the airways.
  (b) True Some people respond to increased ventilation of cold air in this way.
  (c) True Ventilatory capacity falls as resistance increases.
  (d) False The problem is in the bronchial tree, not the alveoli.
  (e) False The O_2-carrying power of the blood may be increased if there is associated polycythaemia.

561 In someone with diabetes mellitus being treated by injected insulin:
  (a) Omission of an insulin injection causes a rise in the blood glucose level
  (b) Regular daily exercise increases insulin requirements
  (c) Unaccustomed exercise leads to a low blood glucose level
  (d) Carbohydrate intake should be decreased if daily exercise is increased
  (e) Tremor may occur if the blood glucose level falls

562 During maximal exercise, a 75 kg athlete aged 25 would have a:
  (a) Heart rate nearer 200 than 150 beats/min
  (b) Stroke volume nearer 80 than 160 ml
  (c) Tidal volume nearer 2 than 1 litre
  (d) Blood lactate nearer 50 than 10 times the resting level
  (e) Mixed venous blood oxygen content nearer 100 than 150 ml/litre

563 Loss of the menstrual cycle (amenorrhoea) in a female athlete may be associated with:
  (a) Loss of the normal monthly rhythms of ovarian hormones
  (b) Weight loss more commonly than with weight gain
  (c) A direct effect on the ovaries rather than on the hypothalamic and pituitary control mechanisms
  (d) Reversal of the condition when strenuous exercise is discontinued
  (e) A body fat level of 25% total body mass

564 The metabolism of:
  (a) Glycogen by exercising muscle leads to a respiratory quotient nearer 0.7 than 0.8
  (b) Fat liberates more than twice the energy liberated by the same weight of carbohydrate
  (c) Fatty acids by skeletal muscle is abnormal
  (d) Amino acids for energy is decreased by cortisol
  (e) An 80 kg male athlete in training requires nearer 2000 kcal (8.4 MJ) than 3000 kcal (12.6 MJ)/day

565 Hypoxia in:
  (a) Exercising muscle decreases the rate of lactate formation
  (b) Life at high altitudes leads to a respiratory acidosis
  (c) Patients with cardiac failure is of the hypoxic variety
  (d) Patients with asthma is alleviated by treatment with beta adrenoceptor-blocking drugs
  (e) Smokers is due partly to carboxyhaemoglobin formation in blood

561 **(a)** **True**   There is impaired uptake and storage of glucose in muscle fibres.
    (b) **False** Exercise facilitates uptake of glucose in muscles and less insulin is needed.
    (c) **True**   The normal insulin dosage is now excessive.
    (d) **False** Exercise increases carbohydrate requirements.
    (e) **True**   It is a sympathetic response to the hypoglycaemia.

562 **(a)** **True**   The predicted rate (220 minus age in years) is 195.
    (b) **False** A heart rate of 195 and stroke volume of 80 would give a cardiac output under 16 litres/min; maximal output in an athlete is about twice that value.
    (c) **True**   A tidal volume of 2 litres and respiratory rate of 60/min gives a total ventilation of 120 litres/min.
    (d) **False** It should be five to ten times the resting level.
    (e) **True**   It would be below 100 ml/litre (less than 50% saturated).

563 **(a)** **True**   The hormonal changes are responsible for menstruation.
    (b) **True**   Moderate weight loss for any reason tends to cause amenorrhoea.
    (c) **False** It is usually a hypothalamic response to a reduction in the body's energy stores.
    (d) **True**   It usually reverses when energy stores, especially fat, are replenished.
    (e) **False** This is a normal female body fat level.

564 **(a)** **False** The respiratory quotient for carbohydrate is 1.0.
    (b) **True**   The ratio is about 9:4.
    (c) **False** Muscle energy is derived from a mixture of carbohydrate and fat.
    (d) **False** Cortisol favours this catabolic process.
    (e) **False** Training greatly increases energy needs, e.g. up to 4000 kcal/day.

565 **(a)** **False** Lack of oxygen leads to anaerobic glycolysis and lactic acid formation.
    (b) **False** It stimulates ventilation and leads to a respiratory alkalosis.
    (c) **False** It is 'stagnant' hypoxia due to inadequate tissue blood flow.
    (d) **False** Beta receptor stimulation leads to relaxation of airway smooth muscle.
    (e) **True**   Due to the carbon monoxide content of the smoke.

566 Exercise at a level of 10 METS:
    (a)   Implies a rate of energy consumption ten times that of the basal metabolic rate
    (b)   Requires an oxygen uptake of 2–3 litres/min in the average adult
    (c)   Requires an oxygen uptake of less than 2 litres/min in a 20 kg child
    (d)   Is not suitable for a person on insulin treatment for diabetes mellitus
    (e)   Is probably too much for a fit 90-year-old person to maintain for 1 h

567 The elastic recoil of muscles and tendons in the legs:
    (a)   Increases jumping height when someone jumps from a height immediately before take off
    (b)   Improves performance during sprinting
    (c)   Contributes more to performance when sprinting on a cinder track than on a concrete surface
    (d)   Can be improved by training
    (e)   Is greater in weight lifters than in skiers

568 The risk of osteoporosis is increased:
    (a)   In older people between 60 and 90 years of age
    (b)   In males as compared with females
    (c)   During prolonged periods of bed rest
    (d)   When both ovaries are removed in a premenopausal woman
    (e)   During treatment with adrenal glucocorticoids

569 During a hospital treadmill exercise test, cardiac abnormality is suggested by:
    (a)   A heart rate greater than 150 beats/min
    (b)   A systolic arterial pressure greater than 150 mmHg
    (c)   ST depression greater than 5 mm in the ECG
    (d)   Inability to follow the usual protocol because of discomfort in the legs
    (e)   A falling systolic arterial pressure during the test

570 For the average healthy, normal male aged 20, the:
    (a)   Body fat is about 30% of total body weight
    (b)   Skin-fold thickness is higher than in a female
    (c)   Heart rate during maximal exertion is about 200 beats/min
    (d)   Cardiac output during maximal exertion is about 10 litres/min
    (e)   Maximal oxygen consumption is about 10 ml/ kg/min

566 (a) False Conventionally it is ten times the resting rate (higher than the basal rate).

(b) True 10 METS require ten times the resting oxygen consumption of around 250 ml/min.

(c) True Resting metabolic rate is roughly proportional to body mass.

(d) False Such people should exercise at this level, but advice is needed on dietary and insulin needs.

(e) True The world record for the mile at this age is between 10 and 15 min, implying less than 10 METS for about a quarter of an hour.

567 (a) True The elastic tissue in extensor muscles is stretched by the initial downward jump.

(b) True Elastic recoil aids the activity independently of muscular contractions.

(c) False The concrete surface 'reflects' more of the energy stored during landing the foot on the surface.

(d) True Training which stretches the muscles achieves this.

(e) False Compared with skiers, weight lifters produce little muscle stretch and rebound during training.

568 (a) True Bone density declines during this period.

(b) False The risk is considerably greater in females, especially after the menopause.

(c) True Normal calcification requires exposure of the bone to gravitational and other stress.

(d) True This leads to an artificial menopause and a considerably increased risk.

(e) True These lead to breakdown of bone collagen, thereby reducing bone strength.

569 (a) False This would be normal in someone under 50 years of age.

(b) False Systolic pressure rises to 200 mmHg in fit young people.

(c) True This level (0.5 mV) indicates inadequate blood flow (ischaemia).

(d) False The abnormality (inadequate flow, muscle fatigue, etc.) is in the legs.

(e) True This means that the heart is failing to pump adequately and is a serious sign.

570 (a) False The normal is around 15–20%.

(b) False Females normally have greater deposits of fat in the skin.

(c) True This is the predicted value (220 minus age in years).

(d) False It is around 20–25 litres/min.

(e) False Average values are around 30–40 ml/kg/min.

571 Heat:
- (a) Load during maximal exertion should not exceed three times resting heat load
- (b) Syncope is caused by an inappropriately high cardiac output
- (c) Stroke is a less serious condition than heat syncope
- (d) Adaptation results in the subject having a smaller rise in core temperature for a given level of work
- (e) Adaptation takes about 6 days rather than 6 weeks to develop

572 Pain is produced by:
- (a) Potassium ions more than by sodium ions
- (b) Occluding the circulation to an exercising limb
- (c) Occluding the circulation to a resting limb
- (d) Raised tissue endorphin levels
- (e) Thawing of tissue frozen during frostbite

573 Heart disease may limit exercise tolerance by:
- (a) Reducing the patient's maximal cardiac output
- (b) Increasing the left ventricular ejection fraction
- (c) Depriving cardiac muscle of an adequate blood supply
- (d) Decreasing heart rate through increased vagal tone
- (e) Changing the relationship between cardiac work and fibre length

574 Respiratory disease can limit exercise by halving the:
- (a) Vital capacity
- (b) Airway conductance
- (c) Body's resting oxygen consumption
- (d) Oxygen-carrying capacity of the blood
- (e) Rate of pulmonary blood flow

575 Olympic level endurance fitness is associated with an exceptionally high:
- (a) Haemoglobin level
- (b) Oxygen saturation of the blood
- (c) Vital capacity
- (d) Cardiac vagal tone during maximal exercise
- (e) Resting stroke volume

571 (a) False   It normally increases 10 to 15-fold in a fit individual.
   (b) False   It is caused by cardiac output not rising sufficiently to compensate for the fall in peripheral resistance due to the skin vasodilation.
   (c) False   Unlike heat syncope, heat stroke is likely to be fatal unless treated promptly and effectively.
   (d) True   Due to the development of more efficient heat-losing mechanisms.
   (e) False   Six weeks gives useful and fairly complete adaptation.

572 (a) True   Injections of isotonic potassium but not sodium into the skin are extremely painful.
   (b) True   Due to the accumulation of pain-producing metabolites when blood flow is inadequate to clear the metabolites generated by muscle exercise.
   (c) False   Anaesthesia due to nerve hypoxia usually occurs before metabolite retention in the tissues rises to the levels needed to stimulate pain nerve endings.
   (d) False   Endorphins inhibit pain pathways.
   (e) True   Thawing releases pain mediators and restores function to numbed nerves.

573 (a) True   This is the main mechanism limiting exercise.
   (b) False   It decreases ejection fraction and hence stroke volume and cardiac output.
   (c) True   This causes temporary (angina) and potentially permanent (myocardial infarction) dysfunction.
   (d) False   It may decrease heart rate by blocking impulses from the sinoatrial node to the ventricles.
   (e) True   It decreases work done at a given fibre length on the left ventricular function curve.

574 (a) True   This may occur in severe obstructive or restrictive disease.
   (b) True   This may occur in obstructive disease such as asthma.
   (c) False   Unless resting oxygen requirements are met, death results immediately.
   (d) False   The oxygen-carrying capacity of blood often rises in respiratory disease because of polycythaemia; desaturation of haemoglobin is the usual cause of hypoxia in respiratory disease.
   (e) False   Pulmonary blood flow tends to rise in respiratory disease.

575 (a) False   Not unless the individual has been training at high altitude.
   (b) False   This also is likely to be normal (around 98% in the arteries)
   (c) False   Ventilation is not usually a limiting factor for endurance activity.
   (d) False   There will only be sympathetic tone to the heart at maximal exercise.
   (e) True   This gives a low resting pulse rate and a high maximal cardiac output.

576 Exercise at high altitudes is hindered by:
  (a) Increased resistance by the atmosphere to athletic activity
  (b) A fall in the total oxygen supply available to muscle during maximal effort
  (c) Respiratory alkalosis
  (d) The compensatory fall in blood bicarbonate level
  (e) High blood lactate levels during severe exertion

577 Malnutrition can limit exercise tolerance by:
  (a) Reducing muscle bulk and hence strength
  (b) Depleting stores of glycogen and fat
  (c) Causing hypoxic rather than anaemic hypoxia
  (d) Causing iron deficiency
  (e) Depleting energy stores even though daily food energy intake exceeds 2500 kcal (10.5 MJ)

578 Increased arousal during competitive sport is indicated by:
  (a) Dilation of the pupils
  (b) A high resting heart rate
  (c) Increased muscle blood flow
  (d) Decreased sweating
  (e) Tremor

579 In someone whose activities are limited by a recent stroke:
  (a) Weakness is more of muscular than neurological origin
  (b) Voluntary movements are better preserved than reflex movements such as the knee jerk
  (c) Speech difficulties are usually worse when the weakness is left-sided
  (d) Weakness usually affects both legs
  (e) The main area of nervous system damage is usually in the cerebellum

580 During strenuous (12 METS) as compared with moderate (6 METS) dynamic exercise, there is a higher:
  (a) Carbon dioxide production in the body
  (b) Systolic arterial blood pressure
  (c) Blood lactate level
  (d) Arterial blood pH
  (e) Respiratory exchange ratio

576 **(a)** **False** Atmospheric drag is reduced at high altitudes.
  (b) **True** Due to impaired saturation of the blood with oxygen (hypoxic hypoxia).
  (c) **True** Due to increased ventilation induced by hypoxia; this limits ability to increase ventilation.
  (d) **False** This compensates for the low P_{CO_2} and allows increased ventilation.
  (e) **True** The lactate levels can rise rapidly and fall slowly, leading to intense fatigue and collapse.

577 **(a)** **True** Due to a shortage of amino acids to repair loss due to normal turnover.
  (b) **True** Thus the substrate for energy production in exercise is missing.
  (c) **False** A reduced haemoglobin level is common and causes anaemic hypoxia.
  (d) **True** This is the most common cause of anaemia (haemoglobin requires iron).
  (e) **True** High activity requires more energy than this so there would be negative energy balance.

578 **(a)** **True** Due to stimulation of radial dilator fibres by sympathetic nerves.
  (b) **True** Another sympathetic effect.
  (c) **True** Due partly to circulating adrenaline.
  (d) **False** Sweating increases as part of the fight or flight response.
  (e) **True** Another beta sympathetic adrenoceptor response.

579 **(a)** **False** The weakness is of neurological origin.
  (b) **False** Spinal reflexes such as the knee jerk are usually increased.
  (c) **False** The muscles on the right side and speech are usually controlled by the left side of the brain; both are usually impaired when the left side of the brain is damaged by a stroke.
  (d) **False** A stroke usually affects just one side of the body.
  (e) **False** The main motor fibres pass from the cerebral cortex to the brain stem via the cerebrum, not the cerebellum.

580 **(a)** **True** Like O_2 consumption, CO_2 production is a precise index of metabolic rate.
  (b) **True** Systolic pressure rises with the level of exercise as cardiac output increases.
  (c) **True** With exercise at 12 times resting metabolic rate, the lactate level is high and rising.
  (d) **False** pH falls due to the lactic acid production.
  (e) **True** This rises as the maximal level (10–15 METS for most people) is approached as buffering of lactic acid leads to CO_2 production.

581 An environmental temperature of 40°C:
   (a) Is thermoneutral if there is a strong wind
   (b) Leads to a generalized release of sympathetic vasoconstrictor tone
   (c) Leads to an increase in heat loss by convection and radiation from the skin
   (d) May lead to heat stroke if relative humidity is 100%
   (e) Is appropriate for people in training for heat adaptation in a climatic chamber

582 A feeling of anxiety before a sporting event may be associated with an increase in:
   (a) Heart rate
   (b) Parasympathetic activity
   (c) Beta adrenoceptor stimulation
   (d) Circulating levels of adrenaline
   (e) Resting respiratory rate

583 An appropriate daily dietary energy intake for a 70 kg adult:
   (a) Who is sedentary and wishes to lose weight is nearer 1500 than 3000 kcal (6.3 versus 12.6 MJ)
   (b) Convalescing from a wasting illness would be nearer 3000 than 2000 kcal (12.6 versus 8.4 MJ)
   (c) In training averaging 10 METS for 6 h/day is more than twice that required for sedentary conditions
   (d) Athlete should be sufficient to avoid ketoacidosis
   (e) Should contain as little carbohydrate as possible

584 The increase in ventilation during maximal exercise:
   (a) May exceed 100 litres/min in a young 70 kg adult
   (b) Is smaller when the exercise is performed at high altitudes
   (c) Is related in part to a fall in pH due to lactic acidosis
   (d) Is related in part to stimulation of receptors in muscle ('metaboreceptors') by metabolic products of exercise
   (e) Does not involve the brain above the level of the brain stem (pons, midbrain and medulla)

585 Glycogen stores in skeletal muscle:
   (a) Are beyond the resolution of the electron microscope
   (b) Help the body to avoid ketoacidosis during exercise
   (c) Increase during endurance training
   (d) Are replaced more completely if carbohydrate is ingested immediately, rather than hours, after exercise
   (e) Are depleted in people with inadequately treated diabetes mellitus

581 (a) **False** The maximal thermoneutral temperature is around 30°C.
   (b) **False** Thermoregulatory release of vasoconstrictor tone in confined to skin blood vessels.
   (c) **False** Since the ambient temperature is higher than blood temperature, the only avenues left for heat loss are evaporation of sweat and fluid lining the respiratory tract.
   (d) **True** Exercise in such environments is inadvisable.
   (e) **True** Provided the relative humidity is low.

582 (a) **True** An increased heart rate is a feature of increased sympathetic drive.
   (b) **False** This tends to produce a relaxed, drowsy state, as after a large meal.
   (c) **True** Beta-blocking drugs reduce the tachycardia and tremor which exacerbate anxiety.
   (d) **True** This produces tachycardia and tremor.
   (e) **True** This is another physical correlate of anxiety.

583 (a) **True** This gives a daily deficit of about 500 kcal, which would be supplied by metabolizing fat stores.
   (b) **True** This provides a positive energy balance and allows replenishment of body protein and fat, assuming an adequate protein content of the diet.
   (c) **True** Metabolizing at 10 METS for 6 h would itself use up more than twice the daily requirement.
   (d) **True** Ketoacidosis impairs optimal muscle function.
   (e) **False** A moderate carbohydrate intake is required to avoid ketoacidosis; a relatively high carbohydrate intake is required for exercise.

584 (a) **True** This is necessary to maintain a normal alveolar oxygen level and full blood oxygen saturation.
   (b) **False** It is increased and ventilatory ability then becomes a limiting factor.
   (c) **True** This occurs above the 'anaerobic/lactate threshold'.
   (d) **True** Ventilation is somewhat reduced if this input to the brain is blocked.
   (e) **False** The precise matching of ventilation to exercise requirements is thought to take place in a postulated 'exercise centre' which receives an input from the cortical neurones initiating the exercise.

585 (a) **False** Glycogen granules are readily identifiable in sections of biopsied muscle.
   (b) **True** By providing an adequate source of glucose.
   (c) **True** This is an important consequence of training; muscle glycogen stores may then exceed 1 kg.
   (d) **True** This has been confirmed in studies involving muscle biopsies.
   (e) **True** Insulin promotes glucose uptake by muscle for replenishment of glycogen stores.

586 This diagram shows two blood oxygen dissociation curves. A represents the oxygen partial pressure in normal alveoli, H the lowered alveolar oxygen pressure in hypoxic lungs due to high altitude or pulmonary disease and V the mixed systemic venous oxygen pressure in the person suffering from hypoxia. In this diagram:

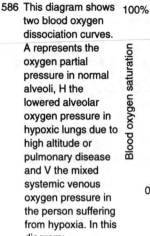

Oxygen partial pressure

(a) If (i) is a normal person's curve, then (ii) is the hypoxic person's curve, rather than vice versa

(b) The blood in curve (i) has a higher red cell level of 2,3-diphosphoglycerate (2,3-DPG)

(c) The O$_2$ saturation of blood leaving the hypoxic lungs is lower than normal in curve (ii)

(d) The oxygen extracted by the tissues is similar for both curves, other things being equal

(e) The curve labelled (i) is more suitable for fetal conditions than the curve labelled (ii)

587 The diagram opposite indicates some events during two respiratory cycles, where I = inspiration and E = expiration. In the second cycle, tidal volume was three times that in the first cycle. Expiration was not forced. It can be concluded that:

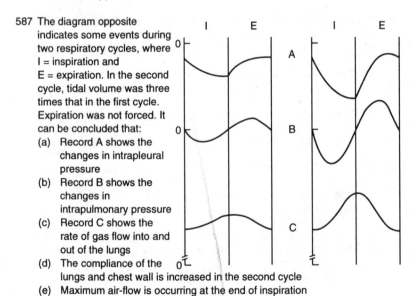

(a) Record A shows the changes in intrapleural pressure

(b) Record B shows the changes in intrapulmonary pressure

(c) Record C shows the rate of gas flow into and out of the lungs

(d) The compliance of the lungs and chest wall is increased in the second cycle

(e) Maximum air-flow is occurring at the end of inspiration

586 **(a) True** The hypoxic person's curve is displaced to the right.

    (b) False 2,3-DPG shifts the curve to the right.

    (c) True This is apparent from the graph and is a disadvantage of the increased 2,3-DPG.

    (d) True XY = YZ, i.e. decreased alveolar uptake of O_2 (X – Y) is compensated by increased O_2 delivery to the tissues (Y – Z).

    (e) True Fetal blood is shifted even further to the left and is able to take up O_2 at the low P_{O_2} levels found in the maternal sinusoids.

587 **(a) True** Intrapleural pressure is negative throughout the cycle and is minimum at the end of inspiration.

    (b) True Intrapulmonary pressure reaches its minimum around mid-inspiration and its maximum around mid-expiration.

    (c) False The flow record closely follows the intrapulmonary pressure record (B) since flow is directly related to the pressure gradient between lungs and atmosphere.

    (d) False Compliance, the volume change for a given pressure change, is similar in both; although the pressure gradient is increased about three times in the second cycle, so is the tidal volume.

    (e) False Air-flow is zero at end-inspiration; it is maximum in mid-inspiration and mid-expiration.

588 In the acid–base diagram below, where L and U represent the lower and upper levels of normal, respectively, a patient whose arterial blood values were found to be at point:
(a)   V might have a compensated metabolic alkalosis
(b)   W might have uncompensated respiratory alkalosis
(c)   X might have a compensated metabolic alkalosis
(d)   Y might have a partly compensated respiratory acidosis
(e)   Z might be suffering from severe vomiting

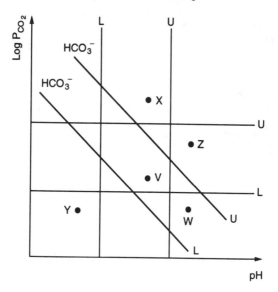

589 The figure below shows simultaneous records of changes in right hand and left forearm volume when collecting pressures are applied intermittently by means of cuffs on the right wrist and left upper arm, respectively. The volume of the hand in the plethysmograph was 300 ml and the volume of the forearm was 600 ml. These records (venous occlusion plethysmograms) show:
(a)   A greater hand blood flow in the third than in the fourth plethysmogram
(b)   That the rate of blood flow per unit volume of tissue is similar in the hand and forearm
(c)   That hand blood flow is more variable than forearm blood flow
(d)   That forearm blood flow is nearer 20 than 50 ml/min
(e)   That the collecting cuffs are not occluding the arteries

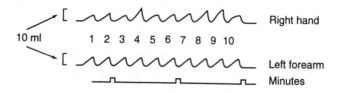

588 (a) **False**   Since all parameters are within normal range, acid–base balance is normal.

(b) **True**   The rise in pH is associated with a low P_{CO_2} but a normal HCO_3^-.

(c) **True**   Or a compensated respiratory acidosis.

(d) **False**   The patient has a partly compensated metabolic acidosis.

(e) **True**   The patient has an uncompensated metabolic alkalosis.

589 (a) **False**   Flow is measured as the increase in volume with time and hence is directly proportional to the slope of the volume record during collection.

(b) **False**   The total blood flow rates are similar on the two sides; since hand volume is half that of the forearm, its rate of flow per unit volume of tissue is approximately double forearm flow.

(c) **True**   Due to greater variability in the resting level of sympathetic vasoconstrictor tone.

(d) **False**   The rate of flow is quite close to 50 ml/min.

(e) **True**   If the arteries were occluded, flow would be zero.

590 Results are given below of a person's vital capacity (VC), forced expiratory
volume in 1 s ($FEV_{1.0}$) and peak flow rate (PFR). The subject of these tests:
(a) is more likely to be a man of 25 than a woman of 65
(b) Is more likely to be suffering from restrictive than obstructive disease of
the lungs
(c) May have asthma, chronic bronchitis or emphysema
(d) Is likely to have an arterial P_{CO_2} 50% above normal
(e) May have a compensated respiratory acidosis

	Observed (O)	Predicted (P)	O/P
VC	4.0	5.3 litres	76%
$FEV_{1.0}$	2.0	4.4 litres	45%
$FEV_{1.0}$/VC%	50%	83%	56%
PFR	200	645 litres/min	31%

591 The diagram below shows left ventricular (LV) function curves of the
Frank–Starling type. If point X on curve B represents the conditions in the
normal heart at rest, then point:
(a) Z might represent conditions in the failing ventricle
(b) Y might represent conditions in the ventricle in hypertension prior to failure
(c) Y, rather than point V, might represent conditions in the ventricle after
administration of a beta adrenoceptor agonist drug
(d) V might represent conditions in a patient with aortic valve stenosis prior to
failure
(e) W might represent the condition in hypovolaemic circulatory failure

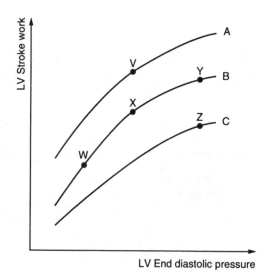

590 **(a) True**   The predicted values are those of a man of 25 (height 70 inches, 1.8 m); for a woman of 65 (63 inches, 1.6 m), the $FEV_{1.0}$ of 2.0 litres would be normal.

(b) False   The relatively severe reductions in $FEV_{1.0}$ and PFR are typical of severe obstructive disease; in restrictive disease, $FEV_{1.0}$ and VC are reduced to a similar extent.

(c) True   All of these produce a similar 'obstructive' pattern of respiratory function.

(d) False   Respiratory failure is a rare complication of obstructive airways disease.

(e) True   If the condition leads to some carbon dioxide retention.

591 **(a) True**   Stroke work is subnormal; end-diastolic pressure increased.

(b) False   In early hypertension the left ventricle hypertrophies; stroke work is greater than normal at a given filling pressure (curve A).

(c) False   Such a drug, e.g. isoproteronol (isoprenaline), mimics sympathetic stimulation and moves the ventricular curve upwards and to the left.

(d) True   Stroke work is considerably increased, end-diastolic pressure little increased.

(e) True   Ventricular function is normal but filling of the heart is inadequate.

592 In this diagram of blood carbon dioxide dissociation curves:

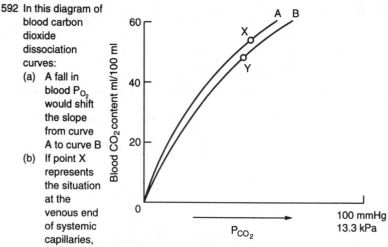

(a) A fall in blood P_{O_2} would shift the slope from curve A to curve B

(b) If point X represents the situation at the venous end of systemic capillaries, then point Y represents the situation of the same blood at the venous end of the pulmonary capillaries

(c) A rise in blood P_{CO_2} would shift the slope from curve B to curve A

(d) The decrease in the slope of the curves as P_{CO_2} rises is related to the saturation of plasma with CO_2 as P_{CO_2} rises

(e) At a P_{CO_2} of 50 mmHg, the amount of CO_2 in solution is lower in curve B than in curve A

593 In the diagram below, the line VXYW represents the threshold of hearing at various frequencies for a normal subject. The:

(a) Sound waves with the characteristics represented by point Z are audible to the subject

(b) Interval AB on the ordinate represents 2.0 rather than 20 dB

(c) Point D on the abscissa corresponds to 5000 rather than 1000 Hz

(d) Segment XY includes the frequencies most important in the auditory perception of speech

(e) Curve is shifted downwards in the presence of background noise

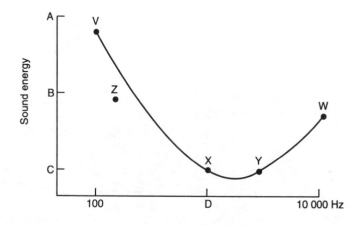

592 (a) **False** Deoxygenated blood can carry more CO_2 than oxygenated blood.
   (b) **True** In the lungs the blood oxygen saturation rises, shifting the blood from curve A to curve B and the CO_2 content falls from point X to point Y.
   (c) **False** It would merely shift the position on a given dissociation curve.
   (d) **False** Plasma does not become saturated with CO_2; CO_2 content remains proportional to P_{CO_2}.
   (e) **False** The amount of CO_2 in solution is the same for both curves for any P_{CO_2}; differences in total CO_2 content are due to differences in bicarbonate and carbamino content.

593 (a) **False** Anything below the line in inaudible.
   (b) **False** AB and BC both represent 20 dB.
   (c) **False** It corresponds to 1000 Hz; the frequency (or pitch) scale is logarithmic.
   (d) **True** Pitch discrimination is best in the range 1000–3000 Hz (XY).
   (e) **False** It is shifted upwards since extraneous (masking) noise raises auditory threshold.

594 In the diagram below, illustrating the handling of glucose by the kidney:
  (a) Line A represents the rate of glucose filtration by the glomeruli
  (b) Line B could represent the rate of absorption of glucose by the proximal convoluted tubules
  (c) Line C follows curve DE rather than angle DFE because different nephrons have different thresholds
  (d) H indicates the maximal reabsorbing capacity of the kidney for glucose
  (e) Renal poisons such as phlorhizin lower the value of H and shift line B to the right

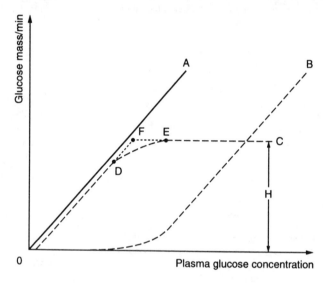

595 In a patient with a red cell count (RCC) of 4×10^{12}/litre, a haemoglobin (Hb) of 7.5 g/100 ml and a haematocrit of 0.28:
  (a) The mean corpuscular haemoglobin (MCH) is nearer 20 picograms (pg) than 20 nanograms (ng). *1 pg = 10^{-12} g; 1 ng = 10^{9} g*
  (b) The mean cell volume (MCV) is nearer 95 than 70 fl (1 femtolitre = 1 μm^3)
  (c) The mean corpuscular haemoglobin concentration (MCHC) is nearer 30 than 35 g/100 ml
  (d) The cause of the anaemia is most likely to be vitamin B_{12} deficiency
  (e) The patient requires a blood transfusion

594 (a) True   The rate of filtration is directly proportional to the plasma concentration.

  (b) False   Line B represents the rate of excretion of glucose; line C represents glucose reabsorbed.

  (c) True   The increased efficiency of glucose reabsorption around the threshold level also contributes to the 'splay'.

  (d) True   The diagram features are typical of a transport system with a limited maximal reabsorbing capacity.

  (e) False   Though H is lowered, the reduction of T_m (glucose) results in glucose excretion at lower plasma glucose levels so the line B shifts to the left.

595 (a) True

$$MCH = \frac{Hb/100 \ ml}{RCC/100 \ ml} = \frac{7.5}{4 \times 10^{11} \ g} = 18.75 \times 10^{-12} \ g = 18.75 \ pg$$

This is below normal (27–32 pg).

  (b) False

$$MCV = \frac{Red \ cell \ volume/litre}{Red \ cell \ count/litre} = \frac{0.28}{4 \times 10^{12}} \ litres = 70 \ fl$$

Since the normal volume is about 75–95 fl, these cells are microcytic.

  (c) True

$$MCHC = \frac{Hb}{Haemocrit} = \frac{7.5}{0.28} = 26.8 \ g/100 \ ml$$

Since the normal MCHV is about 30–35 g/100 ml, these cells are hypochromic.

  (d) False   This microcytic, hypochromic picture is characteristic of iron deficiency.

  (e) False   Moderate iron deficiency anaemia of this sort responds well to iron therapy: blood transfusions should not be used unless absolutely necessary.

596 The diagram below shows two electrocardiogram records from a patient.
Record B was taken 1 year after record A was obtained. In these records:
  (a) The QRS axis in A is directed to the left rather than to the right of vertical
  (b) The QRS complexes V_1, V_4 and V_6 in A suggest left, rather than right,
       ventricular hypertrophy
  (c) The change from A to B suggests a return towards normality
  (d) The QRS axis in B is directed downwards rather than upwards
  (e) The inversion of T in leads III and V_1 in record A indicates myocardial
       ischaemia

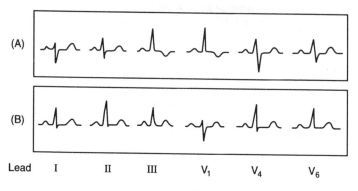

Lead   I   II   III V_1 V_4 V_6

597 The diagram below shows some relationships between lung volume
(increasing upward) and oesophageal pressure during normal tidal breathing.
In this diagram:
  (a) The intraoesophageal pressure is equal to atmospheric pressure at point
       A
  (b) The changes during the respiratory cycle follow the path ABDC
  (c) The slope of the line AD increases when lung compliance increases
  (d) The width of the loop CB increases when airway resistance increases
  (e) AD increases in length during exercise

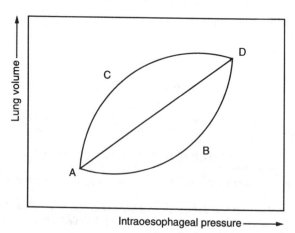

596 (a) False The net QRS in lead I is negative, indicating an axis to the right rather than to the left; this is confirmed by lead III having a large positive deflection.

(b) False The R dominance in V_1 and the prominent S wave in V_4 suggest an abnormally great contribution from the right ventricle in these leads.

(c) True In record B, leads V_1 and V_4 show a normal S wave (V_1) and R wave (V_4), respectively.

(d) True It is roughly 60° below the horizontal (close to the axis of lead II). The swing from right (a) to left (b) indicates return of left ventricular dominance, as do the V lead changes.

(e) False T wave inversion is common and usually has no sinister significance.

597 (a) False Intraoesophageal pressure is similar to intrapleural pressure, which is negative with respect to atmospheric pressure at the beginning of a normal inspiration.

(b) True Since volume changes lag behind pressure changes, the relationship is a hysteresis loop.

(c) True Since compliance is volume change per unit pressure change.

(d) True The greater the airway resistance, the greater the hysteresis.

(e) True Due to the greater tidal volume in exercise.

598 Below is shown the visual field of a normal left eye as plotted by perimetry. When the eye is focused on point Y, an object at point:
- (a) W is detected in the lower nasal quadrant of the left retina
- (b) Y is detected in the region of the fovea of the macula
- (c) Z rather than at point X may be invisible
- (d) W is appreciated as a result of impulses transmitted in the left rather than the right optic tract
- (e) V is seen in monocular vision

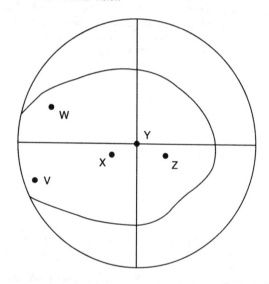

599 Samples taken from the ureters of a man with severe one-sided renal artery stenosis gave the results shown below. Plasma creatinine level was 1 mg/100 ml and the PAH level was 3 mg/100 ml. In this patient:
- (a) Glomerular filtration rate (creatinine clearance) was 10 times as great on the left side as on the right
- (b) Renal plasma flow was approximately 67 ml/min on the left
- (c) Renal blood flow was 900 ml/min on the right (haematocrit = 33%)
- (d) The right kidney had the narrowed renal artery
- (e) It is likely that he was hypertensive

	Left ureter	Right ureter
Urine volume (ml/min)	0.2	6.0
Creatinine conc (mg/100 ml)	100.0	10.0
PAH conc (mg/100 ml)	1000.0	150.0

598 (a) **True**   The image is inverted and reversed with respect to the object.

   (b) **True**   The point focused upon is detected at the macula where visual acuity is greatest.

   (c) **False**   The reverse is the case; the optic disc is medial to the fovea, hence the blind spot is in the temporal part of the field of vision.

   (d) **False**   Impulses from the temporal region of the field of vision cross the midline at the optic chiasma.

   (e) **True**   The visual fields of the two eyes do not overlap for this point.

599 (a) **False**   Creatinine clearance (UV/P) was 20 ml/min on the left and 60 ml/min (normal) on the right.

   (b) **True**   From PAH clearance – a very low value.

   (c) **False**   Blood flow = plasma flow $\times$ 1/1 – Ht = 300 $\times$ 3/2 = 450 ml/min.

   (d) **False**   The left side had the abnormality.

   (e) **True**   Due to renin release from the ischaemic kidney.

600 The diagram below shows results obtained during a glucose tolerance test on
three people. The person represented by curve B was normal. The oral
glucose load was given at time zero. It can be deduced that:
 (a) Curve A is consistent with a diagnosis of diabetes mellitus
 (b) Curve C is more consistent with a diagnosis of an insulin-secreting tumour
     than of malabsorption
 (c) A person showing curve B, who has glucosuria 30 min after glucose
     ingestion, is likely to have a low renal threshold for glucose
 (d) The renal clearance of glucose 2 h after glucose ingestion in patient A is
     nearer 10% than 30% of the renal plasma flow, assuming a normal renal
     threshold
 (e) The renal clearance of glucose for the patient showing curve B is likely to
     be about 60 ml/min 30 min after glucose ingestion

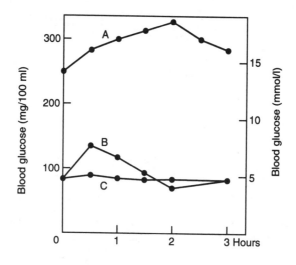

600 **(a) True**   The fasting glucose level and the peak level are markedly raised and there is delayed return of blood glucose to the fasting level.

(b) False   Curve C is typical of the flattening obtained with malabsorption; with an insulin-secreting tumour, the fasting level would tend to be low, with a more marked rise to a peak and a subsequent trough below the initial level.

(c) True   The normal renal glucose threshold is about 180 mg/100 ml (10 mmol/litre).

(d) True   About 150/330 of the filtered glucose is being lost, corresponding to a clearance of 40–50% of the GFR, i.e. about 60 ml, which is about 10% of renal plasma flow.

(e) False   Since blood glucose does not exceed the renal threshold for glucose, renal clearance will be zero.

601 For each paragraph below, select the most appropriate option from the following list of regional blood flow determinants:

1 increased cerebral vascular resistance
2 decreased cerebral vascular resistance
3 increased coronary resistance
4 decreased coronary resistance
5 increased splanchnic resistance
6 decreased splanchnic resistance
7 increased vascular resistance in skin
8 decreased vascular resistance in skin
9 increased renal vascular resistance
10 decreased renal vascular resistance
11 increased regional perfusion pressure
12 decreased regional perfusion pressure

(a) A patient with a head injury receives artificial hyperventilation to reduce cerebral oedema
(b) It has been found that gastric mucosal intracellular acidosis as an indicator of local stagnant hypoxia is useful in assessing splanchnic blood flow in peripheral circulatory failure
(c) Patients with myocardial infarction show electrocardiological improvement after treatment to convert plasminogen into plasmin within 6 h of heart attack
(d) A patient suffering from chronic respiratory failure with carbon dioxide retention has headaches and is found to have papilloedema
(e) An elderly patient suffering from diarrhoea and vomiting for several days cannot sit or stand without developing loss of consciousness (syncope)

601 (a)   Option 1 – *increased cerebral vascular resistance.* Hyperventilation leads to constriction of cerebral vessels due to washout of carbon dioxide from the body. This leads to decreased cerebral capillary pressure and a reduction in cerebral interstitial fluid volume, thereby reducing the oedema generated by head injury.

(b)   Option 5 – *increased splanchnic resistance.* In circulatory failure, blood pressure is supported by increased peripheral vascular resistance induced by the baroreceptor reflex, particularly in the splanchnic circulation. Splanchnic vasoconstriction occurs early in the condition. The resultant stagnant hypoxia in the alimentary mucosa is thus a sensitive index of the early stages of circulatory failure before more severe effects such as hypotension are obvious.

(c)   Option 4 – *decreased coronary resistance.* Myocardial infarction results from complete or almost complete cessation of perfusion of a region of cardiac muscle due to blocking, often by thrombosis, of a coronary artery or arteries. This reduces flow by a massive increase in resistance and a considerable mass of myocardium is threatened by stagnant hypoxia due to poor flow (ischaemia). Activation of circulating plasminogen to plasmin allows breakdown of blood clot (thrombolysis) and decreases regional coronary resistance to a level which allows recovery of ischaemic areas.

(d)   Option 2 – *decreased cerebral vascular resistance.* Carbon dioxide is an important determinant of cerebral blood flow by its local vasodilator action (in underperfused areas carbon dioxide accumulates and this leads to vasodilation and restoration of normal perfusion). When there is a raised level of carbon dioxide in arterial blood there is generalized cerebral vasodilation and this leads to increased formation of tissue fluid (oedema). The resultant increased intracranial pressure leads to headaches and papilloedema, imitating the effects of an intracranial tumour or abscess.

(e)   Option 12 – *decreased regional perfusion pressure.* With persistent diarrhoea and vomiting, extracellular fluid volume can fall severely. The reduced plasma volume leads to hypotension, especially in the elderly whose compensatory mechanisms are blunted. Sitting and standing trap circulating fluid in the feet, causing a severe fall in arterial blood pressure. This decreases cerebral perfusion pressure to a point at which a local decrease in vascular resistance cannot compensate and loss of consciousness (syncope) occurs from cerebral ischaemia (inadequate blood flow).

602 For each case of bladder abnormality, select the most appropriate option from the following list:

1 atonic bladder with overflow
2 stress incontinence
3 chronic prostatic obstruction
4 acute retention of urine
5 automatic bladder
6 bladder diverticulum

(a) A 30-year-old woman with three children complains of wetting herself during coughing and sneezing

(b) A 20-year-old woman had a spinal injury 2 years ago as a result of diving into shallow water. She has lost normal control of the urinary bladder but can initiate micturition when the bladder is fairly full by pressing on the lower abdomen

(c) An 80-year-old man has been admitted to hospital as an emergency complaining of lower abdominal pain and inability to pass urine for 12 h. In recent months he had noticed that the urinary stream was poor. On admission he has abnormal dullness to percussion over his lower abdomen and on rectal examination, enlargement of the prostate

(d) A 29-year-old man was admitted to hospital following a neck injury and paralysis of the legs. On the day after admission, knee and ankle jerks cannot be elicited and he is incontinent of urine

(e) A 75-year-old man complains of frequency of micturition; his cystometrogram shows raised bladder pressures in the contracted state and an abnormally high residual volume

602 (a)   Option 2 – *stress incontinence*. During coughing and sneezing a Valsalva manoeuvre is carried out and this markedly raises intrathoracic and intra-abdominal pressure. In the presence of impaired sphincter action at the bladder outlet, a condition associated with damage at this site during delivery, the raised pressure can expel a small to moderate volume of urine. Laughing may have a similar effect.

(b)   Option 5 – *automatic bladder*. The patient had a spinal injury which has led to loss of bladder control. Such injuries isolate the micturition centre in the sacral cord from higher centre control. In such patients the bladder can empty automatically when distended by means of the bladder stretch reflex centred in the sacral cord. Pressure on the abdomen can initiate the reflex at a convenient time before it occurs automatically.

(c)   Option 4 – *acute retention of urine*. In elderly men prostatic enlargement leads to progressive compression of the prostatic urethra This leads to increasing resistance to flow so that the urinary stream is poor. Often the obstruction becomes complete fairly abruptly so that micturition is impossible and the bladder becomes painfully distended.

(d)   Option 1 – *atonic bladder with overflow*. This is another case of spinal injury isolating the micturition centre in the sacral cord. However, just after such an injury the patient usually shows a complete absence of spinal stretch reflexes for several weeks – spinal shock. Typically the tendon jerks in the region below the injury are absent. The micturition stretch reflex is also abolished so that the bladder loses tone, becomes distended and leaks uncontrollably due to the high pressure in the passively distended organ.

(e)   Option 3 – *chronic prostatic obstruction*. This is another case of prostatic obstruction but without acute retention of urine. Gradual narrowing of the prostatic urethra raises the urethra resistance which the bladder must overcome. Hypertrophy of the bladder wall occurs (as in the left ventricle in systemic hypertension), hence bladder pressure during a micturating cystometrogram (record of bladder pressure versus volume) is increased. Again, as in the failing heart, the bladder muscle fails to empty as completely as usual.

603 For each action or function related to calcium and phosphate metabolism, select the most appropriate option from the following list:
   1   increased blood calcium
   2   decreased blood calcium
   3   increased urinary phosphate excretion
   4   decreased urinary phosphate excretion
   5   increased alimentary absorption of calcium
   6   decreased alimentary absorption of calcium
   7   hydroxylation of cholecalciferol

   (a)   An action of parathormone which increases the likelihood of renal calculi by its effect on renal tubular function
   (b)   A hepatic function which is necessary for normal absorption of calcium from the gut
   (c)   An outcome of the action of parathormone on osteoclasts
   (d)   A renal function which is necessary for the absorption of calcium from the gut
   (e)   A consequence of vitamin D deficiency in childhood which can lead to softening and deformity of bones (rickets) due to inadequate mineral content

603 (a) Option 3 – *increased urinary phosphate excretion*. Parathormone liberates both calcium and phosphate ions from bone and in order not to exceed the solubility product for these ions it is necessary to excrete the excess phosphate. Parathormone favours this by inhibiting reabsorption of filtered phosphate. It thereby tends to raise the solubility product for these ions in urine, favouring development of renal calculi.

(b) Option 7 – *hydroxylation of cholecalciferol*. Cholecalciferol is ingested or synthesized in the skin under the influence of sunlight. To become an active hormone promoting absorption of calcium from the gut it must be converted into 1:25 dihydroxycholecalciferol. The first hydroxylation takes place in the liver.

(c) Option 1 – *increased blood calcium*. Parathormone stimulates osteoclasts to erode bone, thereby releasing calcium and phosphate. This raises the blood calcium level; the phosphate is excreted as discussed above.

(d) Option 7 – *hydroxylation of cholecalciferol*. The second hydroxylation necessary for activating vitamin D (cholecalciferol) takes place in the kidney under the influence of parathormone. Both hydroxylations (liver and kidney) are necessary before vitamin D can regulate total body calcium (mainly in bones) by stimulating its active absorption in the upper small intestine.

(e) Option 6 – *decreased alimentary absorption of calcium*. When vitamin D is deficient (dietary plus lack of adequate sunlight) the substrate for hydroxylation and activation is not available and absorption of calcium is deficient so that there is inadequate calcium for normal bone mineralization.

604 For each case of disordered haemostasis, select the most appropriate option from the following list:

1 capillary abnormality
2 deficiency of factor VIII
3 increased fibrinogen level
4 deficiency of prothrombin
5 deficiency of vitamin K
6 excessive heparin activity
7 massive blood transfusion
8 platelet count 90×10^9/litre
9 platelet count 20×10^9/litre

(a) A 15-year-old child is admitted to hospital with recent onset of widespread purpura (pin-head areas of haemorrhage into the skin). Laboratory investigations reveal an abnormality which accounts for the bleeding tendency

(b) A 50-year-old man is receiving anticoagulant therapy (warfarin, a vitamin K antagonist) after heart valve replacement. He is admitted to hospital with haematuria (blood in the urine) and his INR (international normalized ratio, a measure of the prothrombin clotting time in relation to the normal time) is found to be 4.2

(c) A 90-year-old women has blotchy purple areas about 5 cm in diameter on her hands and arms. They are not uncomfortable and she has no health complaints

(d) A 70-year-old man is operated on for aneurysm (swelling) of his aorta. Severe bleeding requires infusion of 40 units of blood. His recovery is complicated by a bleeding tendency and he is found to have a very low level of fibrinogen. His treatment includes administration of heparin

(e) A 10-year-old child with no known medical problems has been admitted to hospital for persistent bleeding after tooth extraction. Haemostasis had been achieved initially after the extraction but subsequently prolonged oozing from the tooth socket began

604 (a) Option 9 – *platelet count 20 × 10⁹/litre*. Widespread purpura is due to failure of platelet plugging of capillaries and may be due to a low platelet count or to capillary abnormality. An abnormal laboratory test to account for this would be a low platelet count. Although both those given are below normal, only values below 20–40 × 10^9/litre account for serious bleeding.

(b) Option 4 – *deficiency of prothrombin*. The action of warfarin, a vitamin K antagonist, is to impair formation of several coagulation factors, notably prothrombin. There is a number of cardiological indications for the use of warfarin, including heart valve replacement. The value quoted is above the usual recommended range and the prolonged prothrombin time due to a low level of prothrombin would account for the bleeding.

(c) Option 1 – *capillary abnormality*. With advancing age capillaries, like tissues, generally become less resilient in the face of stress such as a relatively high internal pressure. This leads randomly to patchy areas of bleeding such as those described. Apart from their appearance they cause no problems.

(d) Option 7 – *massive blood transfusion*. Massive blood transfusion may lead to widespread activation of the coagulation mechanism – diffuse intravascular coagulopathy. This in turn causes so much deposition of fibrin that the circulating fibrinogen level falls to levels which result in a bleeding tendency. Paradoxically heparin, by preventing the abnormal coagulation, allows the fibrinogen level to rise and can relieve the condition.

(e) Option 2 – *deficiency of factor VIII*. This condition (haemophilia) does not interfere with initial haemostasis due to vascular closure, so the bleeding time is normal as in this case. However, when the vascular spasm wears off, failure to clot is revealed as a persistent ooze of blood. Treatment is by supplying the missing factor VIII.

605 For each case of disturbed acid–base balance, select the most appropriate option from the following list of results of arterial blood analysis:

	pH	P_{O_2} (kPa)	P_{CO_2} (kPa)	HCO_3 (mmol/litre)
1	7.15	16	3	11
2	7.4	14	5	25
3	7.25	9	8	32
4	7.55	10	3	20
5	7.55	11	7	32
6	7.2	25	9	32

100 mmHg = 13.3 kPa

(a) A 60-year-old woman who suffers from long-standing chronic bronchitis has just been admitted to hospital because her condition deteriorated when she developed a chest infection. She is drowsy and cyanosed. No treatment had been given before the blood sample was taken

(b) A 50-year-old man with long-standing chronic bronchitis has been in hospital for several days for treatment of an exacerbation. He is receiving oxygen therapy but his condition is deteriorating

(c) A 50-year-old woman with long-standing renal disease has been admitted with deterioration of her condition, including marked drowsiness. She is noticed to be hyperventilating

(d) A 25-year-old man is taking part in a mountain climbing expedition in the Himalayas and the medical officer of the team is carrying out physiological measurements. The subject has been through the usual protocol for acclimatization to high altitude

(e) A 30-year-old man has been admitted to hospital suffering from abdominal pain and general malaise. He has long-standing upper abdominal pain for which he has been treating himself for some years with quite large amounts of sodium bicarbonate which rapidly relieves the pain. He has begun to get muscle spasms in his hands and feet

605 (a) Option 3 – *This patient has features suggesting respiratory failure.* Drowsiness and cyanosis occur in someone with chronic obstructive airways disease. So we are looking for signs of a respiratory acidosis – low pH due to high carbon dioxide levels and a reduced oxygen level to account for the cyanosis. Only option 3 has these three features. In someone with a long-standing respiratory acidosis the bicarbonate is usually raised as in this case. (For comparison, results in option 2 are all average normal.)

(b) Option 6 – *This patient is very similar to the one above except that he has been receiving oxygen therapy for his hypoxic hypoxia.* Deterioration on oxygen suggests the possibility that complete relief of the hypoxia has resulted in respiratory depression with a rising carbon dioxide level and worsening respiratory acidosis. Results in option 6 confirm this with the very high oxygen pressure which can be produced by breathing oxygen together with a high carbon dioxide level and a dangerously low pH. Correct therapy is to give controlled oxygen at, for example, 24–28% and monitor the blood gases so that the oxygen level is above dangerous levels but the carbon dioxide does not rise dangerously.

(c) Option 1 – *This patient has the symptoms of severe renal failure.* This condition leads to a non-respiratory (or metabolic) acidosis, which is confirmed by the very low bicarbonate level and the very low pH. Such a condition leads to respiratory compensation by hyperventilation to lower the carbon dioxide level as shown. The hyperventilation also raises the oxygen level towards that in the atmosphere.

(d) Option 4 – *High altitudes lead to hyperventilation.* This is triggered by the carotid bodies in response to hypoxic hypoxia. The hyperventilation improves the oxygen level (which is still below that at sea level) but produces a respiratory alkalosis due to washout of carbon dioxide. With acclimatization the kidney responds by lowering the bicarbonate level by reducing tubular secretion of the now scarce hydrogen ions.

(e) Option 5 – *This is now a rather rare cause of metabolic acidosis.* Ingestion of large amounts of sodium bicarbonate relieves ulcer pain by temporarily buffering the gastric acid. However, the bicarbonate is absorbed and can lead to a metabolic alkalosis. Alkalosis increases the binding of available calcium ions in the blood by plasma proteins and can lead to tetany, which usually starts in adults with 'carpo-pedal' spasm. Metabolic alkalosis is compensated by depression of respiration, allowing the carbon dioxide level to rise and balance the increased bicarbonate level. The oxygen pressure tends to fall with the hypoventilation.

606 For each case of coma, select the most appropriate option from the following list:

1 hyperthermia
2 hypothermia
3 respiratory failure
4 hepatic failure
5 renal failure
6 diabetic ketoacidosis
7 hypoglycaemia
8 drug overdose
9 hypothyroidism
10 raised intracranial pressure
11 brain-stem death
12 persistent vegetative state

(a) A 70-year-old woman has been found in her kitchen unconscious when neighbours noted that she had failed to appear as usual for the second day running. On admission to hospital she is noted to have markedly cold skin but this is attributed to the current cold spell and not regarded initially as serious since the clinical thermometer shows only a slightly low body temperature at 35.5°C

(b) A 5-year-old child admitted after a road traffic accident has been deeply unconscious since admission several days ago and is maintained on a ventilator. There is no response of the pupils to light and the corneal response is also absent. Instillation of ice-cold saline into the external auditory meatus produces no eye movements and there is no spontaneous breathing when ventilation is suspended (after pre-oxygenation) to allow the carbon dioxide level in arterial blood to rise to 50% above normal

(c) A 10-year-old child has been admitted in coma with a history of rapid deterioration with vomiting over the previous two days. Recently the child had lost weight and had been passing urine in large amounts more frequently than usual. Hyperventilation is noted. Laboratory investigations show a pH of 7.1 and a bicarbonate level of 9 mmol/litre

(d) A 35-year-old man known to ambulance personnel as a diabetic has been brought unconscious to hospital. His heart rate is 90 beats/min and the pulse is strong, with blood pressure 150/50. Marked sweating is noted

(e) A 30-year-old man was admitted to hospital in an 'intoxicated' state and with a history of vomiting a cupful of blood on the day of admission. He subsequently became drowsy and lapsed into coma, despite transfusion of blood. He is noted to have yellow discoloration of the skin and sclera

606 (a)   Option 2 – *hypothermia.* 'Accidental' hypothermia tends to affect people of all ages when immersed in cold water or exposed on land to low temperatures without adequate insulation; it also occurs in elderly people during cold weather, usually when they have suffered an illness such as a stroke which means they lie, poorly insulated, in a relatively low temperature. Diagnosis requires a low reading thermometer, usually inserted rectally, since a clinical thermometer is shaken down to an initial temperature around 35–36°C and this may be taken as the temperature even though it is 5–10°C below this. The markedly cold skin is an important clue. Treatment of such patients is often unsuccessful since they have the damage produced by hypothermia added to the underlying condition such as a stroke in this case.

(b)   Option 11 – *brain-stem death.* This diagnosis is only made after careful repeated expert examination and after reversible causes such as options 2, 8 and 9 have been excluded. Survival of the brain stem would be indicated by preservation of the brain-stem corneal and pupillary reflexes, by the presence of nystagmus (jerky eye movements) provoked by stimulation of the vestibular system by the icy saline and by spontaneous breathing movements in the presence of a high carbon dioxide level. (The pre-oxygenation fills the functional residual capacity with oxygen to prevent hypoxic damage during removal from the ventilator.)

(c)   Option 6 – *diabetic ketoacidosis.* This is a typical history of childhood-onset of insulin-requiring diabetes mellitus. Lack of insulin's action leads to failure to assimilate absorbed glucose and other nutrients into the body cells, with resulting malnutrition and glycosuria leading to polyuria. In the absence of adequate intracellular glucose energy production relies excessively on fat as a substrate and this leads to ketones and a huge excess of hydrogen ions (severe metabolic acidosis with pH 7.1 and bicarbonate down to about a third of normal due to buffering). This in turn leads to vomiting, coma and hyperventilation to compensate to some degree for the loss of bicarbonate by lowering the carbon dioxide level. The blood glucose level would be 5–10 times normal.

(d)   Option 7 – *hypoglycaemia.* This patient contrasts with the previous one in that he is known to be a diabetic and his blood sugar is below normal. This is strongly suggested by the hyperdynamic circulation – a compensation for hypoglycaemia (ketoacidosis is associated with a weak pulse and circulatory failure). The sweating is also a sympathetic autonomic response to hypoglycaemia. The initial treatment is an intravenous injection of concentrated glucose. If this fails to restore consciousness then other diagnoses must be pursued. Such patients not uncommonly have repeated episodes of unconsciousness due to hypoglycaemia and become known to ambulance personnel.

(e)   Option 4 – *hepatic failure.* The major clue here is the jaundice (yellow discoloration) which suggests a hepatic cause of coma. Hepatic failure often causes in the early stages a state similar to alcoholic intoxication and indeed the two could co-exist as excessive alcoholic consumption is a common cause of hepatic failure. However, in this case vomiting has occurred (common in hepatic failure complicated by portal hypertension and oesophageal varices). It is likely that some of the blood lost will have been absorbed from the gut and digestion and absorption of this high protein load would cause hepatic coma to develop – it is precipitated by the toxic products of protein digestion which cannot be eliminated by the liver .

607 For each case, select the most appropriate option from the following list of results of circulatory measurements:

	Heart rate per min	Stroke volume ml	Pulse pressure mm Hg
1	60	80	40
2	120	35	20
3	180	110	110
4	40	140	50
5	100	70	70
6	240	15	15

(a) A 30-year-old man has been admitted to hospital for minor elective surgery. He is a long-distance runner of national standard. His cardiac shadow is enlarged on chest X-ray and there is concern about his very slow pulse

(b) A 50-year-old woman has been admitted to hospital for thyroid surgery and is found to have signs of severe uncontrolled hyperthyroidism. Her peripheries are warm and moist and her pulse is rapid and bounding

(c) A 40-year-old woman trains regularly for physical fitness but has been concerned recently about chest discomfort, fearing it indicates coronary artery disease. She undertakes cardiological assessment during progressive exercise on a treadmill and the results correspond to the final stage of severity, rarely reached by the patients assessed. At this stage her systolic arterial pressure was 180 mmHg

(d) A 40-year-old woman reports recent episodes of threatened loss of consciousness during exercise and such an episode occurs during treadmill testing in hospital

(e) A 70-year-old man has been admitted to hospital after he collapsed at home and found he could not sit up without feeling he was about to faint. He suffers from epigastric pain treated by a proton pump inhibitor and has recently noticed that his bowel motions are loose and very dark. On admission he is pale and sweating with cold peripheries and his systolic arterial pressure is 80 mmHg even with the foot of the bed raised

607 (a) Option 4 – *These results are typical of a high-level athlete* . There is a normal resting cardiac output of 5.6 litres/min with a very slow pulse rate compensating for a huge resting stroke volume. Such people have relatively large powerful hearts which contrast dramatically with the large weak hearts of patients with cardiac failure, some of whom may suffer from a pathologically slow heart rate which exacerbates their condition.

(b) Option 5 – *Hyperthyroidism leads to a hyperdynamic circulation at rest* . A rapid strong pulse is associated with a high pulse pressure and an increased cardiac output (7 litres/min in her case) – this is required for the increased metabolic rate due to the overactive thyroid. The increased metabolic rate also generates excess heat, hence the sweating.

(c) Option 3 – *This woman is typical of patients with chest pain not due to coronary artery disease.* In association with regular training she is very fit and can exercise to the maximal level used in the treadmill cardiac stress test. At this stage she shows typical findings of a cardiac output of about 20 litres/min and arterial blood pressure 180/70 (high systolic due to powerful ejection by the left ventricle and rapidly falling pressure due to very low peripheral resistance). She has achieved the maximal predicted heart rate (220 minus age in years).

(d) Option 6 – *Syncope or presyncope during exercise can be due to an abnormal ineffective rapid cardiac rhythm (tachycardia)* . As with the previous case the maximal expected rate is 180 and a rate of 240 does not allow adequate filling for a useful stroke volume. Such a low cardiac output (3.6 litres/min) would lead to loss of consciousness during mild to moderate exercise.

(e) Option 2 – *This patient gives a history suggesting peptic ulcer treated by a drug which raises gastric pH to relieve the pain.* The history is strongly suggestive of chronic loss of blood in the faeces (melaena). With an arterial blood pressure of 80/60 he cannot sustain adequate cerebral blood flow in the upright posture. This is because of the reduced venous return experienced by everyone in the upright position. Raising the foot of the bed maximizes venous return. He is suffering from peripheral circulatory failure due to severe blood loss and blood transfusion is urgently indicated.

608 For each case of fluid balance disturbance, select the most appropriate option from the following list:
1 increased total body water
2 decreased total body water
3 increased extracellular fluid
4 decreased extracellular fluid
5 increased interstitial fluid
6 decreased interstitial fluid
7 increased blood volume
8 decreased blood volume
9 increased plasma volume
10 decreased plasma volume

(a) A 20-year-old mentally disturbed patient has refused all food and drink for several days. Urine volume has fallen to around 100 ml in 5 h. Plasma osmolality has risen to 320 mosmol/litre (previously 290 mosmol/litre)

(b) A 50-year-old man has suffered from vomiting and diarrhoea for several days. His peripheries are cold and he has a heart rate of 120 beats/min and an arterial blood pressure of 90/65

(c) A 50-year-old woman is suffering from weakness and mild confusion. She is found to have a plasma sodium level of 125 mmol/litre (normal about 140 mmol/litre) and has a raised level of vasopressin (antidiuretic hormone)

(d) An 80-year-old woman has been admitted to hospital after vomiting blood. Following transfusion of several pints of blood she has become breathless and is found to have an increased jugular venous pressure

(e) A 40-year-old man has been admitted to hospital with full thickness burns of 40% of his body surface. Next day his blood pressure has fallen. A blood test shows a haemoglobin level of 180 g/litre

608 (a) Option 2 – *decreased total body water.* In the absence of any water intake, a person loses a minimum of around 1500 ml/day (500 ml insensible loss from the lungs, 500 ml insensible loss from the skin and 500 ml as the minimum amount of water which can dissolve excreted solid waste products in the urine). A urine volume of 100 ml in 5 h confirms this condition. After several days there will be a water deficit of around 4–5 litres or 10% of total body water, so the osmolality has risen by about 10%. The water deficit is distributed between intracellular and extracellular fluid and oral water would correct the deficit.

(b) Option 4 – *decreased extracellular fluid.* The patient has lost a considerable volume of intestinal secretions. This fluid is isotonic and rich in sodium and chloride, the main extracellular ions. His main depletion is of extracellular fluid and this is confirmed by signs of severe peripheral circulatory failure evidenced by a low arterial blood pressure despite vasoconstriction (cold peripheries) and a rapid heart rate. He urgently needs replenishment of his extracellular fluid by intravenous infusion of isotonic (normal) saline.

(c) Option 1 – *increased total body water.* Inappropriately raised secretion of antidiuretic hormone causes excessive reabsorption of water as fluid passes through the collecting ducts. This dilutes all body fluids as indicated by the low sodium level (osmolality would be correspondingly reduced). The waterlogging of the body cells impairs function and this effect in the brain is manifested by confusion. Restricted water intake would improve the condition.

(d) Option 7 – *increased blood volume.* Replacement of blood loss is urgent in the elderly, but over-transfusion can increase the blood volume above normal. In the elderly there is an increased risk of heart failure and increasing the blood volume can precipitate this so that the heart cannot adequately clear the venous return. The filling pressure of the two sides of the heart increases, causing pulmonary oedema and breathlessness plus increased systemic venous pressure. Diuretic therapy would reduce blood volume by causing excretion of salt and water, thereby lowering extracellular fluid volume.

(e) Option 10 – *decreased plasma volume.* By damaging capillaries, burns cause increased loss of fluid and proteins from the circulation. Loss of protein reduces the oncotic (colloid osmotic) pressure of the blood and fluid passes from the circulation into the interstitial space. The remaining red blood cells become more concentrated, leading to an increased haematocrit and haemoglobin concentration. Blood pressure falls due to the decreased blood volume and decreased venous return. Infusion of plasma would restore the blood volume and blood pressure.